EXPERIENCING THE IMPOSSIBLE

EXPERIENCING THE IMPOSSIBLE

The Science of Magic

Gustav Kuhn

The MIT Press
Cambridge, Massachusetts
London, England

This book was set in Sabon LT Std by Toppan Best-set Premedia Limited. Printed and bound in the United States of America.

Library of Congress Cataloging-in-Publication Data

Names: Kuhn, Gustav, 1974- author.
Title: Experiencing the impossible : the science of magic / Gustav Kuhn.
Description: Cambridge, MA : The MIT Press, [2019] | Includes bibliographical
    references and index"
Identifiers: LCCN 2018024062 | ISBN 9780262039468 (hardcover ; alk. paper)
Subjects: LCSH: Magic.
Classification: LCC BF1611 .K84 2018 | DDC 793.8--dc23 LC record available at
https://lccn.loc.gov/2018024062

10 9 8 7 6 5 4 3 2

I would like to dedicate this book to my wife, Helen, my wonderful children (Amelie, Ella, Joseph, and Mae), and my mum.

CONTENTS

ACKNOWLEDGMENTS

It has taken me just over a year to write this book, but developing the ideas and doing the research has taken much longer. Science is rarely possible in solitude, and it would have been impossible to complete this project on my own. I am extremely grateful for all of the support that I have received while writing the manuscript, as well as the help in developing the ideas and content that led to writing this book. There are so many individuals whom I would like to thank, and I am bound to forget to mention a few by name.

Let me start by thanking all of the people who provided feedback and comments on individual parts of the book. The biggest thanks go to my dear friend and colleague Ron Rensink, whose comments and insights have been immensely valuable. His often brutal edits allowed me to pack more content into fewer pages; his edits mean you don't have to read my ramblings and thus have saved you precious time. I would also like to thank Ron for the continuous guidance and advice throughout much of my academic career, and for his passion, inspiration, and vision, all of which have helped advance the science of magic. I would also like to thank numerous people who have taken the time to read earlier drafts and have provided valuable comments and suggestions: Lindsay Fitzpatrick, Helen Aspland-Kuhn, Vicky Wyatt, Hugo Caffaratti, and Steve Bagienski.

This book is based on research that I carried out over the past fifteen years, as well as that conducted by others. None of my research would have been possible without the help and support of a wide range of scientists, academics, and magicians. They have all contributed to this endeavor, and it's virtually impossible to name them all. Let me start by thanking my PhD supervisor, Zoltan Dienes, who taught me much about science, and Benjamin Tatler and Michael Land, who helped me carry out my first psychological studies on misdirection, which set the foundation for studying the science of magic.

I would also like to thank all of the past and present members of the MAGIC lab (Mind Attention and General Illusion in Cognition), which include postdocs (Cyril Thomas, Hugo Caffaratti), PhD students (Jeniffer Ortega, Lise Lesaffre, Olli Rissanen, Robert Teszka, Steve Bagienski), master's students (Keir Simmons, Nikolas Koutrakis), and countless undergraduate volunteers (Yuxuan Lan, Rhianne Stewart, Max Pitts, Joel Leighton, Nathaniel Ranger-Lunan, Velvetina Lim). All of their hard work has helped establish the scientific knowledge that underpins many of the ideas that I discuss here.

I have been extremely fortunate to collaborate with some of the brightest and most inspirational scientists, most of whom I also consider to be my friends. As such, let me thank Cyril Thomas, Christine Mohr, Geoff Cole, Benjamin Tattler, Benjamin Parris, Christiana Cavina-Pratesi, David Milner, Gianna Cocchini, Alan Kingstone, Pascal Gygax, André Didier-jean, and the people from Abracademy. Likewise, I would like to thank many of the magicians who have helped me gain a deeper understanding of magic (Arthur Roscha, Lee Hathaway, Jim Cellini).

I was extremely fortunate in having the opportunity to write this book while being employed at Goldsmiths, University of London, and I am very grateful for the support I received from the university. I would also like to thank all of the students who took my "Magic and the Mind" module here and elsewhere and who provided feedback on some of the ideas I am about to explore in this book.

I would also like to thank all of the members of the Science of Magic Association, who have helped advance this endeavor and have made this an inspiring and friendly research environment (Jay Olson, Matt Tompkins, Anthony Barnhart).

Let me also thank Bryan Campbell for the lovely title figures, as well as Rebecca Chamberlain and my children Ella and Joe for some of the hand-drawn figures. I would also like to thank the MIT Press and, in particular, Matt Browne for convincing me to write this book.

Most importantly, I would like to thank my wife, Helen Aspland-Kuhn, who has been incredibly supportive of my work, and has continuously helped me clarify my thoughts and ideas, and my wonderful children, who continuously provide new and unique insights into the meaning of magic. I would also like to thank them for all of the delightful inspiration they have provided throughout. Finally, I would like to thank my mum and dad; I am sure my mum would have loved reading and telling others about the book.

① WHAT IS MAGIC?

WHEN I WAS THIRTEEN YEARS OLD, my school friend Arthur Roscha made an egg appear from behind my ear. Arthur's egg trick won't be remembered as one of the world's best magic tricks, but as a young teenage boy, I was pretty impressed: how could an egg simply appear from behind my ear? By that age, I no longer believed in "real" magic and was pretty confident that Arthur didn't have any supernatural powers. Although my rational self knew that I had been tricked by some clever illusion, I simply could not explain the thing I had witnessed with my very own eyes. Deep inside my brain—the dorsolateral prefrontal cortex to be precise—there was a serious conflict between what I thought was possible and what I had just experienced.

Magic is one of the most captivating and enduring forms of entertainment, with magicians all over the world baffling and amazing audiences by creating magical effects. Magic deals with some of the most fundamental philosophical and psychological questions, and yet, unlike many other forms of entertainment, it has received relatively little attention from people outside its sphere. In recent years, psychologists, neuroscientists, and philosophers have started to study magic more systematically, and the science of magic is now a field of its own.[1] Rather than simply speculating about why magic works, we have scientific data that helps explain the psychological mechanism that underpin these wonderful experiences.[2] In this book, we will discuss the latest scientific research on magic, which provides intriguing and often unsettling insights into the mysteries of the human mind. Magic exploits surprising and even counterintuitive psychological principles, and understanding these cognitive processes will challenge your beliefs about your own capabilities. This book will also help you appreciate the complex and almost magical neurological mechanisms that underpin many of our daily activities. Although magic is one of the oldest forms of entertainment, we are only just starting to ask some of the most pivotal questions: What is magic, and why do people endorse magical beliefs? How much of your world do you really perceive? Can you trust what you see and remember? And are you really in charge of your thoughts and actions? We will start by exploring some of these questions before focusing on the psychological principles that magicians exploit. In the final part, we will see how magic is applied to areas outside entertainment and discuss some of the reasons we are so captivated by magic. Before we can answer any of these questions, however, we must take a closer look at the nature of magic.

By now, you may also be wondering how a teenage boy managed to defy the known laws of science and make things appear out of thin air. I'm afraid you may find the answer to this mystery rather disappointing. And once you know how the trick is done, it will never evoke this magical sense of wonder again. Discovering the secret behind magic tricks is often a bit of a letdown. Indeed, a recent study conducted by magician Joshua Jay and a psychology research team led by Dr. Lisa Grimm at the College of New Jersey discovered that most people do not want to find out how a trick is done, which is why I will reveal the secrets only when necessary. In Jay and Grimm's study, participants were shown video clips of magic tricks that were far more impressive than Arthur's. In one clip, for example, the magician made a helicopter appear. The participants were then given a choice between discovering how the trick was done or watching another magic trick. Sixty percent preferred watching another magic trick, whereas only 40 percent wanted to know how the trick was done, illustrating that people are more interested in experiencing mysteries than in solving them.[3] Interestingly, when my student Rhianne Stewart asked people what they liked and disliked about magic, she discovered that although most people don't want to discover the secret, some don't like magic *because* they don't know how it is done.

I am one of the 40 percent who really want to know how a trick is done. And so, Arthur's egg trick started my fascination with magic: how could such a thing happen? At the time, I had no reason to question the authenticity of the egg. Why would it be anything other than real? To the best of my knowledge, all eggs had hard shells, and when you cracked them, they released a messy liquid. There was simply no way that Arthur could have hidden the egg inside his hand or behind my ear without me noticing it. Of course, this reasoning was based on my erroneous assumption that the egg in front of my eyes was real. But although the egg that had just materialized from behind my ear looked like a real egg, it was actually made of soft sponge covered in white fabric. This trick egg could be squashed into a fraction of its size and thus easily hidden inside your hand. I later learned that secretly concealing objects inside your hand is known as *palming*, one of the most powerful methods in the magician's toolkit. Arthur simply palmed the tiny squashed-up sponge egg inside his hand, reached behind my ear, and released the pressure from his hand. The tiny piece of sponge then expanded to its full size. All he had left to do was reveal the egg, which I assumed had appeared from behind my ear.

I had just experienced a truly impossible event, and teenage Arthur was very pleased with himself. But although the trick was indeed pretty cool, I was more fascinated by the way in which such a simple gimmick could trick people. It is fair to say that this sponge egg changed my life. From that day on, I was captivated by the art of magic. Arthur and I borrowed every book on magic from our local library and spent pretty much every minute outside school learning new tricks. (To be honest, we also spent a lot of time during school practicing our tricks, which I like to think explains my poor grades.) After reading all the publicly available books on magic, we discovered magic shops and were gradually introduced into the secretive world of magic, which opened doors to new knowledge and tricks. We attended magic conventions, where some of the world's top magicians would share their thoughts on magic and teach fellow magicians new tricks. Even nowadays, these magic conventions have massive dealer halls where magic traders from around the world sell their newest tricks and gimmicks, and as any magician will tell you, most novice magicians are seduced into spending lots of money on tricks they will never perform. However, one of the main attractions of these magic conventions is simply the chance to hang out with fellow magicians and discuss your thoughts and ideas about magic. You will always find small groups of magicians huddled around a table, each clutching a deck of playing cards, discussing magic into the early hours of the morning.[4]

Magicians often become obsessed with magic and dedicate their entire lives to this mysterious world of deception and illusion. As teenagers, Arthur and I spent pretty much every spare moment of our lives practicing tricks and talking and thinking about magic. My PhD student Olli Rissanen interviewed many of Finland's best magicians to see what it takes to become a professional magician.[5] It turns out that it takes at least ten thousand hours of practice, which coincides with the amount of time that a musician, athlete, or any other performer needs to become a master of their craft.

Although magic involves countless hours of practicing sleight of hand and choreographing magic routines, it is also an intellectual activity. As such, magicians spend a lot of their time thinking and theorizing. Many of the top magicians, such as Juan Tamariz or Ricky Jay, are not only great performers but also have a thorough intellectual understanding of their art. They are able to dissect and explain individual magic tricks in great detail and have powerful insights into how magic works in general.[6] But if you ask such a magician what magic is, most will go blank and struggle to come up with a meaningful definition.[7] Defining magic turns out to be much more difficult than one might think. But thanks to the combined work of magicians, historians, philosophers, and psychologists, we are getting a better understanding of what it is and what it might ultimately be capable of. In this first chapter, we will explore some of the psychological mechanisms that underpin our experience of magic and look at why we all experience magic differently.

IS MAGIC ABOUT TRICKS AND DECEPTION?

I frequently lecture to the public about the psychology of magic. I usually start by performing a few magic tricks, after which I ask members of the audience to come up with their own definition of magic. People's responses vary, but a common suggestion is that magic is all about tricks and deception. Thinking about magic in these terms seems like a good starting point because it's virtually impossible to imagine a magic trick that does not involve deception and trickery. Indeed, many magicians see it as their job to simply fool their audiences.[8] But is it best to define magic in terms of deception? If we think about magic from a psychological perspective, we need to ask ourselves whether deception elicits the same emotional response that you typically experience when you have been successfully fooled.

Performing magic for magicians is notoriously difficult. This is largely because magicians may already know the trick being performed or are familiar with some of the techniques being used, so they know what they should pay attention to. However, not only can magicians fool fellow magicians, they often pride themselves on doing so. In his book, *Designing Miracles*, Darwin Ortiz notes that "whenever a magician describes a magic performance that particularly impressed him, he almost always uses the word 'fooled.' He'll say things like: 'He fooled the hell out of me,' 'I've never been so fooled in my life.'"[9] As a nonmagician, you may not be particularly impressed by such tricks, but magicians love them precisely because they have been fooled. Does this mean we can equate the pleasure we experience when being amazed by a magical illusion to simply being deceived or fooled?

In everyday life, most people are unhappy (or even angry) when they realize that they've been fooled. A secondhand car salesman may use lies and deception to convince you that all of his cars are in perfect condition. You might even walk away feeling pleased with yourself for having purchased a bargain. But this feeling of euphoria quickly turns to disappointment once you realize that you have bought a lemon. The emotional reactions elicited by being tricked into buying a bad car and by seeing an actual lemon magically appear under a magician's cup are very different, which suggests that magic and deception tap into different psychological mechanisms. At the very least, this difference shows that magic is a distinct experience.[10]

Jamy Ian Swiss, a prominent magic theorist, points out that our enjoyment of magic is strangely paradoxical given the negative emotions we generally experience when we are being fooled.[11] Most people who go to see a magic performance know that they are being deceived and lied to, yet they still find this to be a pleasurable experience. At first, this paradox did not make any sense to me, and I spent a lot of time searching for ways to resolve it. However, it turns out that once you look more deeply into our motivation and the function of lying, the paradox resolves itself. It is commonly assumed that all lies are bad and that we don't like being lied to. But both assumptions turn out to be wrong.

Conducting research on lying is rather challenging, in that it's very difficult to discover if someone is lying or not. However, there are several researchers who have tried to establish the frequency with which people lie, and the results are rather startling. For example, a study by James Tyler and colleagues asked university students to get acquainted with one another through a ten-minute conversation, during which the researchers counted the number of times that participants lied. Rather surprisingly, within this relatively short time period, participants lied on average 2.18 times.[12] Similarly, this study asked students to keep a diary and note when they lied. Almost all of the students admitted that they had lied during the week they kept their diary and, on average, had lied to 34 percent of the people with whom they had interacted. Students are not the only individuals who frequently lie. A survey conducted by the *Observer*, one of the United Kingdom's largest newspapers, revealed that 88 percent of respondents admitted to having feigned delight when receiving a bad Christmas present, and 73 percent had told flattering lies about a partner's sexual ability.[13]

If everyone around me is lying, why am I not getting frustrated all the time? And if lying is bad, shouldn't I avoid people who don't tell me the truth? It turns out that this is why there is no paradox regarding magic and deception. There are lots of situations in which we actually prefer living with a lie rather than the truth. For example, if you are asked whether you like your Christmas present, it is far less awkward to tell a fib than to reveal that you really don't like the new jumper. This lack of motivation to discover the actual truth is known as the Ostrich Effect, and it may account for why most affairs will remain undiscovered, even though 40–50 percent of men and women engage in extramarital relations.[14]

Living in a world where everyone always told the truth might actually be rather unpleasant. For example, imagine you have a secret crush on a workmate and would like to ask her out on a date. After weeks of plucking up courage, you walk up to her desk and ask whether she would like to join you for lunch. She is clearly not interested in you, but rather than telling you so directly, she says that she is currently very busy and therefore won't be available. Hopefully, you will take the hint and in your own way conclude that she is less interested in you than you had hoped. Although she lied, the truth would have been far more hurtful. Our social world is very complex and fragile, and these small lies and evasions of the truth are often used as a social lubricant.

Our paradoxical distaste for deception and enjoyment of magic is now beginning to crumble. The nail in the coffin is the fact that we actually prefer to be surrounded by people who are more economical with the truth. For example, introverts lie less than extraverts, yet we typically perceive them as being socially awkward.[15] Indeed, Aldert Vrij has suggested that we find some people to be socially awkward precisely because they don't lie enough.[16]

Now that we have taken a closer look at our psychological reactions toward lies and deception, it becomes apparent that our enjoyment of magic and our experience of deception may not be all that paradoxical. In the final chapter, we will see that current psychological theories help explain how the context of a magic performance allows us to embrace even negative emotions and experience them as aesthetically pleasing. However, for the time being, let me simply say that although deception forms an important part of magic, it involves much more than simply tricking people.

IS MAGIC ABOUT ILLUSION?

When I ask my students to define magic, they often claim that it is all about illusion. Illusions are typically thought of as experiences that do not match with the true state of the world. There is, however, much more to illusions than meets the eye, and we will look more closely at their nature in chapter 5. But for the time being, this definition is sufficient.

The figure below depicts two horizontal lines. Which of these appears longer?

For most people, the top line looks substantially longer than the bottom one, even though the two lines are physically identical. This is known as the Müller-Lyer Illusion, one of the most studied illusions in vision science. Among its many interesting properties is the fact that the strength of the Müller-Lyer Illusion is not affected by knowing that it is present. These properties are the reason that magicians frequently use such illusions to create their magic effects. However, the experience evoked by magic differs from simply experiencing a sensory illusion.

To see this more clearly, consider one of my favorite visual illusions, Roger Shepard's Turning the Tables.[17] Look at figure 1.2 and decide which of the two tables will fit more easily through a narrow doorway. Most people will select the one on the left because it appears much narrower than the one on the right. However, both table surfaces are identical.[18] As in the case of the Müller-Lyer Illusion, Turning the Tables is very effective even when you know that it is an illusion. Even though the rational self tells you that the surfaces are identical, you simply cannot perceive them as such. In the real three-dimensional world, the table surfaces would certainly be different, and your visual system is generalizing your knowledge

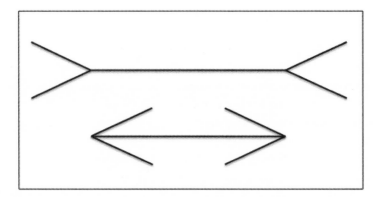

Figure 1.1

Müller-Lyer Illusion

Figure 1.2
Roger Shepard's Turning the Tables

about real-world objects to influence how you perceive two-dimensional shapes.

We will learn much more about the relevant perceptual processes in subsequent chapters, but for now, what is important is the fact that even when you know such illusions are present, the mismatch with reality is still there. Indeed, me telling you about these illusions turns a simple drawing into something much more intriguing. As philosopher Immanuel Kant already noted, "This mental game with sensory illusion is very pleasant and entertaining."[19] Likewise, most people remain intrigued even once they know how illusions are accomplished. Magician Thomas Fraps suggests that our enjoyment of visual illusions, despite knowing how they are accomplished, differentiates magic from illusions.[20] Magic is far more fragile: in contrast to the case of the Müller-Lyer Illusion, once you know that the egg appearing from behind your ear is a gimmick, the enchantment of the magic trick evaporates.

The same point can be made by noting that—as we will learn later on—your perceptual awareness of the world and even of yourself is an illusion. So if everything you experience is effectively an illusion, and if illusion is the basis of magic, then why don't we continually have magical experiences during our day-to-day lives? Even though illusion can play an important role in magic tricks, magic is about much more than illusion.

IS MAGIC ABOUT SPECIAL EFFECTS?

Another possible characterization of magic is in terms of special effects, such as those encountered in cinema. In the early days of cinema, there was a close link between magic and filmmaking; during the Victorian era, magicians such as John Nevil Maskelyne or Howard Thurston regularly incorporated short film screenings into their magic shows.[21] Similarly, French illusionist and film director Georges Méliès combined magic and film to create some of the earliest special effects, and many of his achievements paved the way for modern cinema.

More recently, computer-generated special effects have given Marvel's superheroes their superpowers and have enabled Harry Potter to perform wonders that most magicians can only dream of. Such effects look so realistic that it's often difficult to distinguish them from reality. But even though I have no idea how these effects are achieved, I do not experience them as magic; at the back of my mind, I'm aware that computer algorithms can conjure up pretty much anything on the screen in front of me. Consequently, there is a plausible—although not necessarily correct—explanation of what you are seeing, in contrast to what happens during a magic performance.[22]

An interesting illustration of this point can be found in a set of videos that create a strong magical experience even though no magician—at least, as traditionally construed—is present. In one such video, a man pours water into a glass, covers the glass with a piece of cardboard, and then turns the glass over while pressing the cardboard firmly onto the opening. He then places the upside-down glass on the table. Up to this point, it's all just physics. The man then lifts the cup while making a quick twisting motion with his hand, and the water becomes suspended in the cup's shape. You can clearly see a wobbly, jelly-like structure that immediately returns to its liquid state when touched.[23]

The first time I watched this video, I was very surprised. How is this possible? Doesn't this defy all the known laws of physics? I was pretty sure that it must be a trick, but I could not think of a magic method that could create such a clean and beautiful illusion. The wobbly video was shot using a mobile phone camera, and the production value appeared to be pretty low. Surely, if the cameraman couldn't hold the camera steady, let alone focus correctly, an amateur must have shot the footage. Because the video appeared to have been shot by an amateur, I did not suspect any fancy special effects, and based on the comments, neither did most of the other people who watched it.

It turns out, however, that the video was actually created using state-of-the-art editing techniques and was deliberately made to appear as if it had been shot by an amateur. In essence, the water trick illusion is fooling you twice. First, your perceptual system is fooled by the computer-generated special effects. But more importantly, you are also fooled into believing that the video is real. As such, you experience something that is apparently impossible, a situation very similar to that experienced when watching a magic performance.

IS MAGIC ABOUT THE SUPERNATURAL?

As a magician and scientist, I'm often asked whether I believe in "real" magic, that is, forces beyond those known to current physics. This turns out to be a rather tough question to answer, because yes, I have seen a lot of things that I simply cannot explain. For example, I have no idea how a touch screen allows me to swipe from one page to the next, nor do I have a clue as to how each time I press a key on my keyboard, a letter appears on the monitor in front of me. Many of the technological devices we use today appear magical, but I know that they were created by very clever engineers using technology that is based on modern physics.[24] I therefore don't experience them as magic. Interestingly, there are phenomena that even the brightest minds struggle to understand. For example, how do your conscious thoughts interact with your physical brain? Philosophers and scientists alike struggle to explain this phenomenon, which on its face seems to be inexplicable in terms of current physics. Oddly, however, such phenomena are typically not considered to be magic.[25] We will look at people's beliefs about magic in much more detail in chapter 3, but let us now examine the role of supernatural powers in magic.

Any demonstration that appears to contradict our current understanding of science is generally considered to be magic. Indeed, the online *Oxford Dictionary* defines "magic" as "the power of apparently influencing events by using mysterious or supernatural forces."[26] We might, therefore, try to define magic—at least as performed by magicians—as the emulation of real supernatural powers.

Many magicians use tricks and deception to create the illusion that they have supernatural powers, but they are generally considered to be charlatans. For example, some priests in ancient Greece and Egypt used sophisticated theatrical illusions to create the impression that they were in direct contact with the deities.[27] For example, a priest might light a fire on an altar, after which a set of heavy stone doors would mysteriously open (see figure 1.3). These illusions provided powerful proof of the priest's supernatural powers. But the true force behind this mysterious phenomenon was a clever array of counterweights and expansion tanks: A fire heated the water in a tank located underneath the altar, forcing it into an

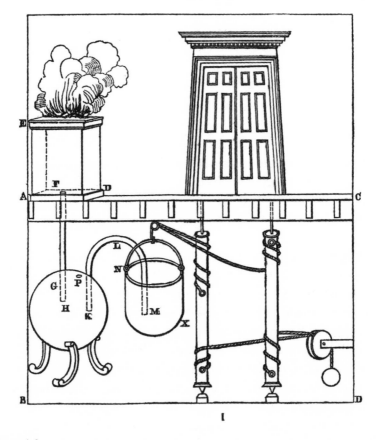

Figure 1.3

Ancient hydraulic mechanism that "magically" opens temple doors, as described by Heron of Alexandria (ca. 10–70 CE)

expansion tank. Once filled, the weight of the water inside the tank would pull open the doors. This clever mechanism was completely hidden, so witnessing the immensely heavy stone doors mysteriously open must have been an awe-inspiring phenomenon.

Similarly, many mediums today use magic tricks to create the illusion that they possess supernatural powers.[28] For example, in 1974, Uri Geller caused a storm when he claimed that he could bend metal objects via the power of his mind and could read people's thoughts. Some of these demonstrations were reported in one of the most respected scientific journals, *Nature*, but none of the claims stood up to scrutiny.[29] It's likely that Geller accomplished these feats by simply bending the objects with his hands and secretly peeking inside sealed envelopes.[30]

Most magicians, on the other hand, do not claim to have supernatural powers, and within the magic community, there has been a strong drive toward exposing those who make such assertions. In the 1920s, magician and escapologist Harry Houdini adamantly confronted people who claimed to have supernatural powers, and he spent much of his career exposing spiritual mediums.[31] Indeed, magicians such as James Randi and Derren Brown have tried to expose Geller and others like him, and Brown has publicly explained how psychics use linguistic tricks and cold reading to give you the false impression that they know specific details about your life.[32]

Although magicians explicitly deny that they have supernatural powers, most magic performances evoke a sense of supernatural deception. For example, although Houdini adamantly attacked anyone who claimed to have spiritual powers, he was happy for people to believe that his super strength and skill enabled him to pick any lock and escape from any prison.[33] Houdini was very proactive in creating this image of a man with near-supernatural powers, and although he was certainly skilled at picking locks, there was more to his escapes than meets the eye. Each of these stunts was a carefully choreographed performance that required meticulous planning, including the bribing of insiders, which created the illusion of Houdini's ability to escape from anywhere.

Adamantly skeptical about any spiritual magic, Derren Brown claims that his illusions are created through clever psychological tricks. In one of his performances, Brown invites two members of an advertising company to a designated location, where they are asked to design a poster promoting a chain of stores.[34] Before they start, he casually places a sealed envelope containing some of his own ideas in front of the men, who then come up with several drafts of their poster. To everyone's surprise, Brown's prediction is a good match. Most traditional magicians would leave it there, but Brown takes the illusion one step further by explaining that the advertisers had been unconsciously primed with that concept before they entered the room. We then revisit their taxi journey, which clearly depicts the two men driving past prominent objects that would later appear in their poster. This is presented as proof that Brown had purposefully primed the men to come up with that particular advertising concept.

Most people I talk to are convinced by Brown's psychological explanations and don't consider him to be a traditional magician. Without giving away the secret, I can assure you that the illusion described above did not involve priming or any other subtle form of mind control. (In chapter 7, we will examine the science behind this form of mind control in more detail.) This effect was created using more traditional magic methods. The real illusion involves you believing that these psychological effects are possible; Brown is honest about the tricks used by spiritualists, but he misleads you about his own psychological powers. The problem he faces is that mysterious psychological effects make his performances interesting and engaging. Thus, he endorses a new form of nonscientific phenomenon even as he dismisses the occult. Brown has stated that he struggles with this paradox, but it's a paradox that all magicians face: magic is only magical if the audience experiences the effect as mysterious, surprising, or supernatural.[35] Although magicians explicitly deny these supernatural explanations, our experience of enchantment nevertheless depends upon those explanations.

IS MAGIC ABOUT SUSPENSION OF DISBELIEF?

When I ask magicians to define magic, they often talk about "suspending your disbelief."[36] Suspension of disbelief is a concept that is often used in theater, which might explain why magicians relate to it. For example, every time you go to the theater, you accept the scenes being acted out on stage as real. You are fully aware that an actor on stage is simply pretending to cry, but by suspending your disbelief, you can enter this world of fantasy. Likewise, when Peter Pan is lifted by steel cables above the stage, you suspend your disbelief and imagine that he is truly flying, regardless of whether or not you can see the cables.

Although a play may appear magical, it is a different type of experience from that encountered when the magician David Copperfield flies across the stage. Seeing strings attached to Copperfield would destroy the illusion, because you are no longer experiencing the impossible.[37] In fact, even suspecting the existence of strings destroys the illusion, and this is why Copperfield flies through hoops and is locked inside a transparent box, while still apparently defying the laws of gravity. Good magic happens regardless of whether or not you are willing to suspend your disbelief. This is why Teller—the silent guy from Penn & Teller—argues that experiencing magic is an unwilling, rather than a willing, suspension of disbelief. According to him, magicians force you to suspend your disbelief, regardless of whether or not you want to do so.[38]

Although we are getting closer to understanding what magic is, the suspension of disbelief idea does not fully capture people's experience of witnessing a magic trick. When I see David Copperfield flying across the stage, I don't truly believe that he can actually fly. In fact, if I believed that people can fly, I would no longer experience this performance as magical.[39] Surely then, magic is about experiencing things that we think are impossible, or what Darwin Ortiz refers to as the "illusion of the impossible."[40]

IS MAGIC ABOUT CONFLICT BETWEEN BELIEF SYSTEMS?

Jason Leddington, a young philosopher at Bucknell University, suggests that suspension of disbelief is not the key to magic. Rather, "the audience should actively disbelieve that what they are apparently witnessing is possible."[41] In other words, magic is only successful if people simultaneously believe and disbelieve what they are seeing. In this view, magic creates a conflict between what Darwin Ortiz calls our "intellectual belief" and our "emotional belief." Your intellectual belief tells you that magic is impossible, but on a more primitive emotional level, the performance induces a belief that magic is actually happening.[42]

Leddington connects this with a philosophical idea proposed by Tamar Szabó Gendler: the concept of *alief*. An alief is an automatic, primitive attitude that may conflict with a person's explicit beliefs. Whereas most of you will be familiar with beliefs, the concept of alief may need a bit more elaboration. Let us consider the following example: In 2007, the Grand Canyon Skywalk opened on the Hualapai Indian Reservation, promising a sensation that, until then, one could only experience in dreams. One tourism website promises that "dreams and reality will meld into one," which immediately conjures up expectations of shamanistic experiences of all sorts. However, this dreamlike experience could not be further from spirituality. The Skywalk simply consists of a horseshoe-shaped glass walkway that protrudes seventy feet beyond the canyon's rim, thus offering stunning views of this marvel of nature. It allows you to stand over the canyon and look down through the crystal-clear transparent floor, leaving you with the feeling of being freely suspended more than two thousand feet in the air.

Leddington suggests that walking onto this viewing platform elicits a feeling that is comparable to experiencing a magic trick because, like magic, it creates a tension between our beliefs and our aliefs. The Grand Canyon Skywalk is safely secured, and the glass has been heavily reinforced and is thus extremely unlikely to break. Even though thousands of people have ventured onto this platform before, the physical process of leaving the hard ground of the canyon rim feels uncomfortable. Indeed, you can see people clinging to the railing, and they hesitate before stepping onto the transparent part of the walkway. The strange sensation that you feel when stepping out onto such a platform is created by the tension between two competing beliefs: "Although the venturesome souls wholeheartedly *believe* that the walkway is completely safe, they also *alieve* something very different. The alief has roughly the following content: 'Really high up, long long way down. Not a safe place to be! Get off!'"[43]

According to Leddington, the experience of magic results from a similar cognitive process. The audience knows that magic is impossible, yet a good magic performance simultaneously induces the belief on a more primitive and emotional level (alief) that the impossible is indeed happening. Leddington's philosophical theory has only just been published, but it has already attracted much interest from magicians, philosophers, and psychologists.

IS MAGIC ABOUT IMPOSSIBILITY?

Most theories of magic assume that it involves experiencing something impossible. Indeed, I have titled this book *Experiencing the Impossible*, as this feels like an intuitive way to capture the emotional sensation that magic elicits. But what does it mean for something to be impossible? Exploring this perspective may give us some additional clues about the nature of magic.

The term "impossible" refers to something that is not possible. This differs from something that is extremely unlikely. It is extremely unlikely that I will win the lottery, but there is still a slim chance that this could happen. On the other hand, most people would consider it impossible for me to levitate my laptop—gravitational forces pull objects to the surface, and without a counterforce (e.g., magnetism), objects simply cannot levitate. Logically speaking, the term "impossible" does not allow for degrees, any more than does "contradictory." And yet, some things seem to be more impossible than others. How could this be?

Walt Disney was one of the world's most successful cartoonists. In his early days, his cartoons were rather surreal, depicting a world where anything was possible: a sausage could jump from the grill and dance in a kick line. However, Disney's more successful feature-length films followed specific magical rules whereby magic needed to be "plausibly impossible," meaning that it could violate some real-world expectations but not too many.[44] For instance, in the movie *Snow White and the Seven Dwarfs*, it is plausibly impossible for forest animals to communicate with Snow White but *implausibly* impossible for them to double in size or to ooze through keyholes.

Andrew Shtulman suggests that in a fictional narrative where everything is possible, we still distinguish between events that are flat-out impossible and those that are impossible but plausible.[45] Shtulman points out a few examples. In *Star Wars—Episode V: The Empire Strikes Back*, the Jedi master Yoda teaches his apprentice, Luke Skywalker, to levitate stones before teaching him how to levitate an entire starship. Likewise, in the magical world of *Harry Potter*, the potions instructor Severus Snape teaches his students how to brew a forgetfulness potion before he teaches them how create a potion for endurance. Similarly, in the fictional world of Disney's *Cinderella*, the fairy godmother turns a pumpkin into a stagecoach and a horse into a coachman rather than turning a horse into a stagecoach and a pumpkin into a coachman. In the real world, rocks are lighter than starships, your ability to bear prolonged hardship requires more mental effort than simply forgetting something, and pumpkins show more resemblance to coaches than to coachmen. These perceptual and conceptual considerations influence your judgments in the real world, yet they should not do so in the world of magic.

Shtulman asked people to review the curriculum from *Harry Potter*'s Hogwarts School of Witchcraft and Wizardry and evaluate the difficulty of various pairs of spells (e.g., levitating a bowling ball versus levitating a basketball). In each pair, both spells violated the same causal principle (i.e., gravity) but differed in terms of their subsidiary principle (e.g., weight). Interestingly, spells that violated deep-seated causal principles and seemed implausibly impossible were rated as harder to create than those that were considered to be plausibly impossible. In other words, even in a world where nothing is impossible, some things are perceived as being more impossible than others.

DOES MAGIC HAVE A NEURAL BASIS?

Now that we have dealt with some of the philosophical and psychological perspectives relating to magic, let us turn to neuroscience. Nearly a decade ago, Benjamin Parris and I tried to discover the parts of our brain that are involved in experiencing magic.[46] This was at a time when scientists still got excited by demonstrating how certain parts of your brain light up when you have a particular thought or carry out a particular task, and we both joked that we could call this theoretical brain region the "magic spot."

We didn't have any official research funding for our quest, and we were therefore forced to turn my bedroom into a temporary film studio, where we recorded different magic tricks. We played video clips of these tricks for our participants while measuring their brain activation using functional magnetic resonance imaging (fMRI) to scan their brains (not in my bedroom). The participants also viewed clips of me doing things with the same objects that did not involve magic but were still surprising. This provided us with a baseline to which we could compare our participants' brain activation.

The results took us both by surprise. Even though we had both joked about hoping to find the magic spot, neither of us seriously believed that magic could be localized to one discrete area because magic involves many different types of experiences. However, to our amazement, once the large amounts of data were processed, we found that two brain areas became particularly active at the moment when participants viewed the magic tricks: the left dorsolateral prefrontal cortex and an area known as the anterior cingulate cortex. The dorsolateral prefrontal cortex is involved in monitoring conflict. Likewise, the anterior cingulate cortex is a part of the brain that tries to resolve conflict. To our surprise, magic activates those parts of the brain that are typically involved in processing and resolving more general types of cognitive conflict.

Our paper reporting these findings, which we titled "Imaging the Impossible," was the first neuroscientific study to explore the neural mechanisms underlying our experience of magic. Since then, others have conducted similar studies and have come to similar conclusions.[47] Then, as now, we did not believe that this or any other neuroscientific study would tell us what magic is. However, it is interesting that observing magic and dealing with other types of conflict both activate the same neurons. This

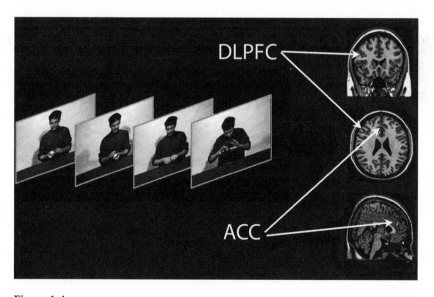

Figure 1.4

Brain activation when people watch a magic trick (adapted from Parris et al., "Imaging the Impossible")

indirectly supports the idea that magic involves a conflict between what we believe to be possible and what we believe we have seen.

MAGIC AS A KIND OF WONDER

At this point, I would like to sum up what we have learned so far about magic. We have learned that magic creates a cognitive conflict between things we experience and things we believe to be impossible. But it is not simply about experiencing things that are impossible; instead, it is about experiencing things we *believe* to be impossible. More precisely, my colleagues and I propose that magic is the experience of wonder that results from perceiving an apparently impossible event.[48]

In this view, at the center of the magical experience lies a cognitive conflict, and the stronger the conflict, the stronger the experience of magic. There are three important things that are worth pointing out. First, magic depends not on what you have actually seen but on what you believe you have seen. This might seem like a subtle distinction, but most techniques in magic involve not simply creating an event but manipulating your beliefs about that event. (We will discuss this in more detail in the next chapter.) Second, because the cognitive conflict also depends on things that we believe to be impossible, rather than things that actually are impossible, people with different beliefs will experience magic differently. Put simply, a mind-reading demonstration will elicit wonder only if you do not believe in telepathy. Third, because we are interested in what people believe to be impossible, we need not restrict ourselves to logical characterizations of impossibility; you can have a stronger or weaker belief in an event's impossibility regardless of whether or not it is actually impossible. Thus, magic can vary in its intensity. For example, although the laws of physics dictate that it is impossible for any object to just disappear, I intuitively believe that it is more impossible for a larger object to disappear than a smaller one; a disappearing elephant will create a stronger magical experience for me than a disappearing mouse. Only time will tell whether this framework holds up, but for the moment, I believe that it can explain much of the experimental evidence collected to date. Let us now look at how this framework helps explain why we all experience magic differently.

DO WE ALL EXPERIENCE THE SAME MAGIC?

If magic relies on a conflict between what we believe is possible and what we perceive, people who hold different beliefs could experience the same magic performance in very different ways. For example, young children tend to blur the boundaries between reality and fantasy and often believe in ideas and concepts that we adults dismiss as being impossible—for example, magical creatures, such as Santa Claus and the Tooth Fairy, and the idea that you can truly make things invisible.[49] Jay Olson, a PhD student at McGill University, recently conducted a study on how children interpret magic tricks.[50] He filmed a simple trick in which a magician used sleight of hand to make a pen disappear. In the video, the magician held a pen between his hands and suddenly appeared to break it in half; when he opened his hands, the pen had vanished. The secret behind this trick is that the pen was quickly moved inside the magician's jacket. Olson wanted to see how children of different ages would explain the trick; he even provided clues as to how it was done. Anyone watching carefully would notice an object hitting the magician's shirt, which clearly hinted at the method.

Olson and his fellow researchers played this video clip to nearly 170 children between the ages of four and thirteen years and asked them to explain what they saw. Younger children (four to eight years old) typically took the magician's actions at face value and claimed that the pen "just disappeared" or simply "dissolved in the magician's hands." Older children (seven to nine years old) developed possible yet still implausible explanations of the trick. For example, although the magician had rolled up his sleeves, many suggested that he had hidden the pen up his sleeves or in his skin. My favorite explanation was that the magician's torso was actually a mannequin and that the magician hid the pen inside the empty torso. These findings demonstrate that as children discover more about the world and learn to distinguish between appearance and reality, they start to interpret magic tricks in less supernatural ways.

Conversely, there are magic tricks that will amaze adults but leave young children cold. This is because adults generally make different assumptions about the world. When we see an object occluded by another object, we typically assume that the first object continues to exist and retains its physical properties; we fully understand that objects continue to exist even though we may not see them. This principle, which the Swiss developmental psychologist Jean Piaget called *object permanency*, allows us to experience a stable world—a world in which objects stick around even when they are out of sight.[51]

Much of magic involves exploiting assumptions we make about the world, and there are countless magic tricks that exploit object permanency.[52] For example, in the French Drop, one of the most common sleights of hand used to make a coin disappear, the magician pretends to transfer a coin from one hand to the other, when in fact the coin remains secretly concealed in the original hand (see figure 1.5). After this simulated transfer, the magician pretends to hold the coin and, even though you cannot see it, you are convinced that it must still be there. After a few moments, the magician opens his hand to reveal that the coin is no longer there—it has vanished!

This simple yet extremely effective coin trick relies on your sense of object permanency. Thus, even as object permanency allows you to experience a stable world, it is also responsible for fooling you into believing that a coin can simply vanish. Most importantly though, it allows you to experience a conflict between your beliefs (i.e., the coin is still in the hand, even though I can't see it) and your experience (i.e., the coin is no longer there).

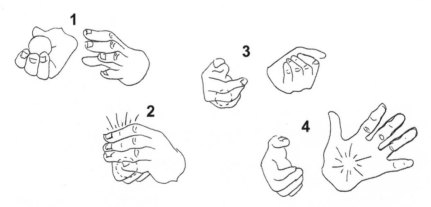

Figure 1.5

French Drop: Sleight of hand, used to vanish a coin. The spectator assumes that the coin has been transferred from the magician's right hand into the left hand, when in reality it remains secretly concealed in the right hand.

How then would a young child respond to this trick? To answer this question, we must look at young children's beliefs about the world. We are born into a very confusing world that is full of overwhelming stimulation and information. Much of our early development involves gradually learning to make sense of the things that are happening around us. Piaget suggested that babies spend their first two years learning about the existence of objects, and he argued that before this step is reached, children consider objects that are out of sight to be nonexistent. There is currently no consensus as to when exactly object permanency develops; although Piaget initially argued that this does not take place until eight to twelve months of age, others have suggested that even fourteen-week-old infants may have some form of object permanency.[53] Even though there is still much disagreement as to how and when object permanency develops, it is clear that young infants' ability to represent and reason about unseen objects is significantly limited and that our sense of object permanency develops through infancy and childhood.

It is this weaker sense of object permanency that explains why babies all around the world are fascinated by peekaboo. To play this game, all you need to do is hide your face for a few seconds and then simply pop back into view and say, "Peekaboo!" As an adult with a fully developed sense of object permanency, we may find it hard to appreciate why babies are so drawn to this game. Early theories of why babies enjoy this game so much assumed that they are simply surprised when the face returns to reality.[54] It was thought that peekaboo creates surprise, similar to the element of surprise in a joke or a magic trick. However, it turns out that there is more to peekaboo than simple surprise because infants don't enjoy the game if you include too much magic. W. Gerrod Parrott and Henry Gleitman have shown that babies get less enjoyment from the game if you surreptitiously change the person who reappears.[55] Adding too much surprise seems to kill the game, and it is thus thought that peekaboo helps babies test and retest the fundamental principle of object permanency. In other words, this simple game may teach babies that things may stick around even though they are out of sight. Given that young children have different beliefs about the world, we can assume that they will experience magic tricks differently than we do.

Piaget conducted most of his early research on his own kids. He gave them puzzles and observed the types of mistakes that they made at different stages in their lives. Given our shared Swiss heritage, I was very keen to emulate Piaget by running experiments on my own children, and many of the experiments involved studying how they responded to my magic tricks. My older children were often amazed when I vanished objects using the French Drop, but my youngest daughter, Mae, who was one year old at the time, was far from impressed. Because babies can't talk, it's difficult to ask them what they think of a trick, but you can gauge their experience simply by monitoring their emotional reaction. It's fair to say that Mae was far more engaged by peekaboo, whereas most of my sleight of hand magic left her pretty cold. This, of course, is not that surprising, because at the time, Mae's sense of object permanency was still developing and, for her, the existence of objects that were out of sight was far hazier. Simply making something disappear would not result in a strong cognitive conflict between what she saw and what she believed about the world, and so she would not experience the magic.

It goes without saying that experiencing a magic trick relies on the spectator not discovering the true method behind the trick; once you have discovered the secret, the cognitive conflict disappears. One of the magician's primary goals is to avoid anyone noticing how the trick is done, which was why I spent nearly every free minute of my teenage years practicing sleight of hand and ensuring that nobody would ever notice when I palmed a coin. Many of my friends and colleagues often comment on how much my children must enjoy having a dad who can perform magic, and being a magician certainly comes with some parenting perks. For example, my children get very excited when I magically pull sweets and coins from behind their ears. Because I have spent a large part of my life dedicated to perfecting my legerdemain, I have always felt that my conjuring skills are superior to those of most mums and dads. However, I soon realized that I was deluding myself. One day, when my oldest daughter, Ella, was four, she came back from nursery school telling me that her teacher was also a magician. I knew that he was not a proper magician, but in my daughter's eyes, there was very little difference between Ed's simple magic tricks and my sleight of hand illusions. Had I lost all of my skills and truly misspent my entire youth?

Just because a child has been fooled by a trick does not necessarily mean that adults won't discover the secret. In fact, adults often discover the secret behind the magic tricks performed by children's entertainers, and most children's tricks will fail to amaze a skeptical adult audience. This is not because children's entertainers are bad magicians; it's simply because their tricks are designed to be enjoyed by children rather than adults. Why then are children more easily deceived? We have run experiments in which we measured participants' eye movements to study differences in the attentional strategies that children deploy when they see a magic trick.[56] We used a specially designed magic trick in which I vanished a lighter by using misdirection. We will discuss this type of magic trick in much more detail later on, but the method behind this trick uses misdirection to draw attention away from the lighter, which enabled me to drop the lighter into my lap. Although this happened in full view, the misdirection prevented most people from noticing the secret. We asked adults and children under the age of ten to watch a video clip of this trick while we monitored where they were looking using an eye tracker. Our results showed that children had far less control over where they were looking and were much more easily misdirected. These results further illustrate that our experience of magic tricks is very subjective and that children and adults can clearly experience the same performance very differently.

Because all of us have very different beliefs about what is possible, we will all experience magic slightly differently. However, inasmuch as magic is about experiencing more general cognitive conflict, it should be possible for nonhumans to experience similar emotions. In recent years, there have been several viral YouTube videos showing how animals react to magic tricks. For example, Jose Ahonen, a Finnish magician, gained a worldwide audience through a video clip entitled "Magic for Dogs," in which he uses sleight of hand to vanish dog biscuits.[57] Once the dogs realize that the biscuits have vanished, they appear genuinely puzzled and confused, similar to how humans react to magical illusions. By now the internet is full of videos of magic tricks being performed for all sorts of animals, and even though it is difficult to know what these animals are experiencing, their reactions suggest that we may not necessarily be alone in enjoying these enchanting illusions.

Magicians have astonished people with their illusions for thousands of years, and they can explain the methods used to create the illusions in much detail. There are thousands of magic books that provide intricate details about how these tricks are performed, but far less is known about the experience they elicit. The combined efforts of magicians, psychologists, philosophers, and neuroscientists are helping to shed light onto this experience. Magic is an experience of wonder that results from a cognitive conflict between things we experience and things we believe to be impossible. However, because we all hold different beliefs and experience the world differently, each magic trick will elicit a somewhat unique emotional experience. Let us now take a closer look at how and why our brains are so easily tricked into believing that we have experienced something that we believe to be impossible.

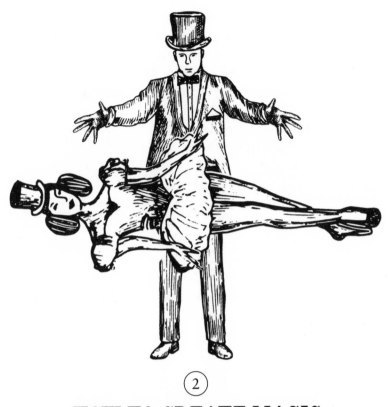

(2)

HOW TO CREATE MAGIC

BACK IN THE 1840s, Parisians talked enthusiastically about ether, a new chemical substance that seemed to have almost magical powers. Jean-Eugène Robert-Houdin, a French magician generally regarded as the grandfather of modern magic, took advantage of this. At the beginning of his illusion, he proclaimed that "if one has a living person inhale this liquid when it is at its highest degree of concentration, the body of the patient for a few moments becomes as light as a balloon."[1] He then placed three stools on a wooden bench and asked his youngest son, Eugène, to stand on the middle one with arms extended. Next, he placed a cane on top of each stool and positioned them under Eugène's arms. Robert-Houdin then produced a small bottle from his pocket and opened it to unleash the pungent smell of ether, which permeated the theater. He placed the bottle under his son's nose, and Eugène immediately went limp. Knocking someone out with ether is not magic; indeed, the chemical was once widely used as an anesthetic. But Eugène had not been sedated. The bottle was in fact empty, and the pungent smell spreading throughout the theater was produced by another son pouring ether onto a hot iron shovel.

Thus far, the Parisian audience had not seen anything extraordinary, but this was about to change. Robert-Houdin removed the stool below his son's feet, leaving him hanging on the canes like a limp rag. He then removed one of the canes so that Eugène was dangling by one arm. The audience must have been amazed. But Robert-Houdin took the illusion a step further: he lifted the boy upright and moved him into a horizontal position using only his little finger. When he let go, Eugène was suspended in the air, balancing on his elbow with no other support. After the audience had feasted on this stunning illusion, and as the ether's effects were starting to wear off, Robert-Houdin returned his son to his original position, at which point Eugène woke up, seemingly none the worse for wear.

Many Parisians at the time believed that the ether had caused the boy to levitate; indeed, Robert-Houdin received countless letters complaining that he was putting his son's health in jeopardy.[2] The illusion, of course, was not created through ether. It instead relied on an ingenious mechanical construction that exploited several weaknesses in the audience's perceptual reasoning.

Many spiritual healers, psychics, and holy men claim that their miracles are created through genuine magic. It is not my intention here to evaluate whether these claims are true (though I have yet to see convincing evidence that supports them). I will say that there are far more reliable and impressive ways to create magical miracles. A professional magician's tricks must always work, and relying on divine intervention is simply not a good way of doing this. Instead, a magician must exploit the limitations of human cognition. In this chapter, we will explore the psychology behind the tricks that make you experience things that you believe are impossible.

PSEUDOEXPLANATIONS

In the absence of divine intervention, how can we create a magical experience? Darwin Ortiz, a magician and magic theorist, suggests that "magic can only be established by a process of elimination."[3] According to him, you conclude that something is magic only because you cannot come up with any other explanation. When Parisians witnessed Eugène levitating, they could not explain how this was being done, which then resulted in a magical experience. According to Ortiz, "The primary task in giving someone the experience of witnessing magic is to eliminate every other possible cause. ... If it can't have been caused by anything else, it must be magic!"[4]

Magic never happens without some cause or other. In *Harry Potter*, magic defies all laws of physics and so seems completely impossible. But at Hogwarts, Harry and his friends spend hours learning about spells and potions, and there is a very explicit causal link between magic spells and magical effects. A particular spell will always turn a toad into a stone, whereas another with always make objects levitate. In Harry's world, it is clear that the magic spell is what causes the toad to turn into a stone.

In the real world, magic spells rarely cause actual transformations. But magicians almost never create a magical illusion without invoking some cause—some form of "magic stuff." Magic stuff is an essential ingredient for any strong magic trick; without it, the effect is merely a puzzle that elicits curiosity but not necessarily wonder. Robert-Houdin went to great lengths to turn an earthly mechanism (i.e., ether) into magic stuff. Today, more than 150 years after its invention, variants of the Ethereal Levitation Illusion are still commonly performed around the world. Come rain or shine, it's nearly impossible to walk through Covent Garden in London without passing someone in a *Star Wars* Jedi costume levitating a few feet above the pavement and hustling passersby for a selfie in exchange for a few coins. These illusions make for amusing Facebook pictures, but they will never elicit the type of reaction that Parisians experienced back in the 1840s because a key ingredient is missing: the Jedi's levitation is not perceived as resulting from magic stuff. Indeed, Robert-Houdin himself suggested that magic should "induce the audience to attribute the effect produced to any cause rather than the real one."[5] Put more simply, the

Figure 2.1

Robert-Houdin's Ethereal Levitation

purpose of magic is not for the audience to experience a cause without an effect but to experience an apparent impossibility, with magic stuff being the only remaining explanation.

The term "magic stuff" may seem rather vague and fluffy, which is why magicians prefer to call it a *pseudoexplanation*.[6] Pseudoexplanations lie at the heart of magic; they are what the magician wants the audience to believe is the true source of the effect. Several kinds of pseudoexplanation are possible. For example, in a typical Pick a Card Trick, you pick a card, and the magician discovers your card's identity through some magical means. Assume that I want to demonstrate that I can read your mind (this is known as *mentalism*). You pick a random card and look at it, and I then ask you to visualize the card in your mind. After some magical gestures and a very intense stare, I pretend to enter your deepest thoughts and then reveal the name of the card you had visualized. The pseudoexplanation here is my ability to read your thoughts via telepathy; the actual method is that the cards are secretly marked, allowing me to identify your card by deciphering the subtle code on its back. The same illusion can be created using other methods. For example, I could *force* your card, a technique by which magicians systematically make you pick a card without your being aware of this. Once I force you to choose the six of spades, I will know your card's identity without having to enter your mind. Alternatively, I could glimpse the chosen card using sleight of hand or a secret device such as a mirror and then pretend to read your mind.

Whereas straightforward mind reading was very popular in the past, many magicians today like giving their performance a bit of a psychological twist and so instead claim that they are reading your body language.[7] For example, I might ask you a series of questions and instruct you to lie about some of the answers. I claim that I am monitoring your eye movements and using neurolinguistic programming (NLP) to determine whether you are lying or telling the truth. To everyone's amazement, I discover your card's identity by reading your body language.

NLP refers to a wide range of psychological techniques that allegedly teach people about the relationship between eye movements and thoughts.[8] However, even though NLP is commonly used in business seminars, there is about as much scientific evidence for the link between eye movements and lying as there is for telepathy or ether's antigravitational effects.[9] In short, NLP is merely a pseudoexplanation—impossible yet still believable.

Another common pseudoexplanation is the claim that the magician has extraordinary mental skills, such as super memory. Instead of claiming to read your mind or monitoring your eye movements, for example, I might look through the deck of cards pretending to memorize them all and then triumphantly reveal that the six of spades is missing. Unlike NLP or telepathy, memorizing an entire deck is possible. Indeed, there is a World Memory Championship, in which Speed Cards is one category. In December 2017, Zou Lujian set a new world record by memorizing a deck of playing cards in 13.96 seconds.[10] Memorizing an entire deck in a short amount of time is therefore not impossible, but it is still pretty challenging. Likewise, magicians may give you the impression that they have the calculating skill of a computer or the strength of an ox. Technically, such illusions are not impossible, they are just extremely unlikely.[11]

In many instances, magicians rely on pseudoexplanations that are far subtler than the examples given thus far. For example, when a magician vanishes a coin, he usually makes a magical gesture, such as snapping his fingers. This gesture is perceived as having caused the effect. Most people watching magic don't intellectually believe that the pseudoexplanation is real; we don't truly believe that magicians have real magical powers. In fact, believing a pseudoexplanation to be real is a way of abolishing the cognitive conflict between belief and experience.[12] The role of the magician is to create a scenario in which you believe that the pseudoexplanation is the only possible cause of the event, even though you're fully aware that it's not real. But if we know that the explanation is not real, why is our mind so willing to accept it?

WHY WE ACCEPT MAGIC STUFF

The magician's main objective is to get the audience to experience a "sequence of events that involves an extraordinary or supernatural cause-effect relation," resulting in a sense of wonder about what just happened.[13] This, of course, is a reductionist view; good magic also includes interesting stories, comedy, and other artistic considerations. For the time being, however, we will focus on what it takes to experience something we believe to be impossible.

Getting people to believe an illusory cause is easier than you might think. The reason for this is that causal relationships must be inferred; you cannot see a cause directly. The British empiricist David Hume noted that inasmuch as there is no metaphysical glue that binds events together, we can never discover the real cause of things.[14] According to Hume, all we can do is notice regularities, such as B will follow A, and then inductively infer a causal relationship. For example, every time in the past that you let go of an object, it fell to the ground. But this does not mean that it will necessarily happen again in the future. It may be extremely likely to happen again sometime, but given that causes cannot be directly observed, there is simply no way to prove that it will always happen.

The human mind, however, seeks to organize our world exactly in terms of cause and effect. Back in the 1940s, Albert Michotte, a Belgian experimental psychologist, showed that even simple visual sequences can appear to be causally connected.[15] For example, imagine a circle and a square sitting on a line, separated from each other by a few inches (see figure 2.2). The circle starts to move in a straight line until it reaches the square, after which the circle stops and the square starts to move along the same trajectory. Strictly speaking, the circle and the square move independently; the stopping of the circle has no implications for the

Figure 2.2

Michotte's launching paradigm

movement of the square. Nevertheless, you experience the circle as causing the square to move. We automatically infer causality between events
even when they occur independently; we do this despite knowing that the
circle does not really cause the square to move.[16] Even though it is unclear
whether or not this causal perceptual reasoning is innate, it is clear that it
develops at a very early age (six months).[17]

Our mind likewise attributes animacy and social goals whenever possible, even for simple geometric shapes. In their influential paper, Fritz
Heider and Marianne Simmel show that people spontaneously describe
simple moving shapes in terms of purposeful actions.[18] Almost all observers describe one shape (a small triangle) as shy and another (a large triangle) as a bully, and they often describe them in terms of attributes such as
anger or frustration. This tendency to infer social intentions from movement patterns develops very early in childhood. For example, Philippe
Rochat and colleagues report that even three-month-old infants prefer to look at displays in which two circles appear to be chasing each
other than at displays in which the circles are randomly moving around
the screen.[19]

Our minds are continuously inferring causality between events, and
in most instances, this causality is simply an illusion; it does not capture
what is really going on. For example, each time you watch a cartoon
on TV, you experience the characters as talking. When Kermit the Frog
opening his mouth coincides with a speech sound, we automatically infer
that Kermit is talking. This illusion—the linchpin of ventriloquism—
is powerful enough to override the true nature of the perceptual signal. It works because our mind assumes that the speech must originate
from the puppet; even though we know that the speech comes from a
different location and that puppets can't talk, we still perceive Kermit
as talking.[20]

Although our minds are continuously interpreting our environments in terms of causal relationships, some relationships are endorsed more readily than others. In a set of ingenious experiments, Eugene Subbotsky investigated the type of explanations that adults endorse after witnessing an impossible event.[21] The experimenter produced a postage stamp and a small wooden box and then asked each participant to place the stamp inside the box and close the lid. The experimenter then pronounced the magic words "As hasher nor hashilym, ud hasher nar uzdalyk," after which he opened the box to reveal that the stamp had been cut in half. Most participants were surprised by this, but even though they did not know how the trick was done, only three of the sixteen participants believed that the damage had been caused by the spell. The experimenter then asked the participant to place their driver's license inside the box and said, "If I repeat my magic spell, I cannot give you the guarantee that your driving license will come out of the box in the same condition that you put it in." Only two of the sixteen participants—all adults, remember—then asked the experimenter not to cast the spell, suggesting that only a few thought it might just work. What happens if we change the nature of the magical causal explanation? In a different condition, Subbotsky used the same magic trick, but instead of the magic spell, he pressed a button on an unknown physical device that produced light and sound effects when it was switched on. There was no direct connection between this device and the wooden box, and there was no logical way in which the device could have cut an object in half. However, now half of the participants spontaneously acknowledged that the action had caused the destruction of the stamp even though none of them could say how this had occurred.

Rationally, there is no difference between reading your mind through supernatural spiritual power and applying NLP techniques to monitoring your eye movements. Similarly, there should be no logical difference between casting a spell and using some fancy physical device to destroy an object that has been sealed inside a wooden box. However, even though we do not know how the process works, we are much more willing to endorse a plausible explanation, such as an invisible physical force, than magic.

CREATING MAGIC THROUGH MISDIRECTION

People are happy to accept an illusory explanation of an effect even when it has no relation at all to what actually happened. How do you get people to believe in such magical causation? Much of this is due to *misdirection*.

Many magicians have attempted to organize and categorize magic tricks so that commonalities can be found. For example, Robert-Houdin divided conjuring into six branches, including feats of dexterity, experiments in natural magic, and mental conjuring.[22] Alternately, Dariel Fizkee claimed that all magic tricks can be broken down into nineteen basic effects, and Norman Triplett came up with an even larger list.[23] All of these systems focus on the magic effects rather than the methods. In a book entitled *Conjurers' Psychological Secrets*, on the other hand, the magician S. H. Sharpe presents a comprehensive list of the principles underlying magic methods.[24] However, rather than simply discussing each of these, I would like to discuss magic method more generally and explore the psychological principles that underpin it. My ultimate—though challenging and probably impossible—aim is to create a framework that explains how all magic works. I'm aware that this is a bold attempt and one that will raise much criticism, but I hope that it will at least encourage a constructive debate.

Back in 2006, I was fortunate to gain a fellowship that allowed me to spend time at the University of British Columbia, where I had the pleasure of meeting Ronald Rensink. Although Ron had no practical background in magic, he was very interested in using it to study human cognition, and his sharp, critical, and open mind proved extremely valuable for this endeavor. Together we have been working on developing a science of magic based on established psychological principles.[25] We are not the first to do so; others, such as Richard Wiseman and Peter Lamont, have already proposed a framework.[26] But the aim of our endeavor was to build a direct bridge between magic and the brain.

At the time, we thought that most magic tricks could be explained through three key principles: illusions, forcing, and misdirection.[27] This approach inspired much scientific research. Now, however, I believe that all magic can be explained through misdirection alone. I cannot take credit for this idea, which was suggested by the magician John Hugard back in 1960, when he wrote that "magic is misdirection and misdirection is magic."[28] Understanding how magic works therefore requires an understanding of misdirection.

Although hundreds of books and articles have been written on the topic of misdirection, a clear understanding of this concept remains elusive.[29] The situation is not helped by the use of the word "misdirection" in many domains outside magic, such as politics. Nor is it helped by the common confusion between misdirection and distraction. To better define this elusive concept, let us start by discussing some commonly held misconceptions about misdirection.

A popular source of information, Wikipedia states that "misdirection is a form of deception in which the attention of an audience is focused on one thing in order to distract its attention from another."[30] This definition highlights two commonly held misconceptions about misdirection: it involves distraction, and it serves to divert attention away from something. As we will learn in more detail in the next chapter, attention plays a critical role in shaping what we see, and attentional distraction can effectively render you blind. For example, while driving, failure to attend to the road ahead can prevent you from seeing a car that is suddenly pulling out in front of you. In the context of magic, however, attentional distraction needs to be handled carefully. I recently saw a magic show in which five clowns chaotically ran around the arena performing ten different magic tricks simultaneously. My attention was successfully distracted, in that I failed to see the methods. But the confusion also distracted my mind from the effects, so that I also failed to experience the magic.

I also encountered this phenomenon in some of my earliest studies on misdirection.[31] My colleagues and I used eye-tracking technology to measure people's eye movements while they watched a simple trick. We wanted to study attentional processes in a natural environment, and because most of our participants were students, the obvious choice was the student bar. We packed up all of our equipment and set up a temporary lab in the bar. However, I had failed to anticipate that most students would be far more interested in our eye-tracking equipment than my magic tricks. Although this provided the perfect distraction to prevent them from noticing the method, it also prevented them from noticing the magic. Misdirection rarely involves simply distracting people's attention; it instead aims to guide attention toward the effect. This is why Tommy Wonder, a true legend in magic, has suggested that the term "misdirection" is misleading and might even misdirect us from understanding misdirection.[32]

Another misconception is that misdirection is simply about preventing you from noticing something. Although there are many ways to prevent you from noticing the secret method used in a magic trick, simply failing to see the method does not guarantee that you'll experience the magical effect. Let me explain using an anecdote: It was a very proud moment when my son Joe performed one of his first magic tricks. Standing in front of me and showing that both of his hands were empty, Joe then asked me to close my eyes and used this "misdirection" to prevent me from seeing that he was secretly removing a pen he had hidden in his back pocket. He then instructed me to open my eyes, and after a flamboyant "ta-da," he made the pen appear right in front of my eyes. Even though I did not see how the pen had appeared, the trick simply did not elicit a strong sense of wonder. (Although it did elicit the cuddles of a proud dad.)

I did not experience Joe's trick as magic because I knew that closing my eyes had prevented me from seeing things, and so rather than accepting Joe's magic as the cause of the effect, I assumed it was the visual interruption. Successful misdirection relies on the audience being unaware that they have been misdirected; at no point should they question their ability to detect the secret. But misdirection is more than this. These tricks only work if people are unaware of the true cause: their own perceptual and cognitive limitations. As soon as we become aware of these limitations, we start to attribute the anomalous effect to our own cognitive failures, rather than magic stuff.

Luckily for the magician, many of our cognitive quirks and inadequacies are counterintuitive. In addition, people tend to resist the idea that they are less than perfect. This gives conjurers ample opportunity to convince people of magic stuff. For example (as we will see later on), even though we believe we are observant, we are aware of only a small fraction of what goes on around us. Once I can control your attention, I can prevent you from noticing things that are right in front of your eyes.[33] These effects are similar to you deliberately closing your eyes, but the critical difference is your belief about what caused the effect. Had Joe misdirected my attention, rather than instructed me to close my eyes, I would have attributed the effect to his magic stuff and not my perceptual interruption.

The manipulation of attention is certainly important for misdirection. I am very fortunate to have trained with a master of misdirection, Jim Cellini, who was one of Slydini's students. Slydini is known as the king of misdirection, as most of his magic involved beautifully choreographed movements that manipulated the audience's focus of attention.[34] His sleight of hand techniques were rudimentary. But there was no need for fancy moves because he perfectly orchestrated people's attention, giving him full control over what they saw. But having said all that, it is important to keep in mind that there is much more to misdirection than simply controlling people's attention.

Many principles in misdirection involve manipulating cognitive processes other than attention, such as how we remember an event or make sense of it. Juan Tamariz is widely regarded as one of the greatest magicians of our time. Among his many contributions to the field of magic are his insights into how magicians misdirect people's memories. As he elegantly puts it, "A magician can create *lagoons* in the spectators' memories in order to make them forget whatever we wish for the magical effect, or to make them believe they remember things that in reality never existed."[35] Many misdirection techniques rely on getting you to falsely remember events, as this is an extremely effective way to prevent you from becoming aware of their true cause. We will discuss these principles in more detail later on.

There are also countless misdirection techniques that involve manipulating the way you reason about an event. For example, the *theory of false solution* is a misdirection principle that can be used to influence our memory or our reasoning.[36] According to this principle, the magician presents the audience with an obvious yet false solution that is later revealed to be wrong.[37] An extreme form of this can be found in "sucker tricks," such as the Egg Bag routine. In the Egg Bag Trick, an egg appears and disappears inside a cloth bag. The magician then pretends to sneak the egg under his arm, after which he shows the bag to be empty. The real method involves a secret compartment inside the bag that allows the magician to conceal the egg. But given that the audience suspects that the egg is under the magician's arm, they are less likely to discover the true explanation, even after their suspected solution is shown to be false.

Cyril Thomas, an excellent magician who consistently baffles me with the rubber band tricks he's always inventing, has conducted intriguing research that is providing insights into some of the cognitive mechanisms that underpin the theory of false solution.[38] Cyril and I have recently shown that the theory of false solution truly prevents people from discovering the correct solution.[39] In this experiment, Cyril claimed that he would make a card magically move from the deck he was holding in his hand to his back pocket. He simply showed observers the six of clubs, made a magical gesture, and then pulled another six of clubs out of his pocket. This is not a particularly amazing trick, and nearly 80 percent of the participants correctly deduced that he had pulled a duplicate card from his pocket. However, once he added a false solution, the trick became much more impressive. The false solution involved pretending to palm the six of clubs from the top of the pack, but he destroyed this as a potential solution by revealing that his hand was empty before it reached the pocket. Participants knew that palming was not a possible solution, but they still struggled to discover the real solution. This false solution not only hindered participants from discovering the correct solution, but it also made this trick seem significantly more impressive. People are reluctant to abandon a false solution; they tend to fixate on a false solution even when they know that it is wrong, which prevents them from considering an alternative. Psychologists refer to this phenomenon as the Einstellung effect,[40] and it is related to functional fixedness, wherein people can only use an object in the way it is traditionally used.

Now that we have highlighted several aspects of misdirection, let's try for a definition: misdirection is an umbrella term describing a range of psychological principles that prevent the audience from discovering the true method and focus the audience's attention on the effect. As a magician, I think of misdirection as a process that guides you toward accepting the "magical" explanation as the only possible one, in the most natural way possible. As a psychologist, I think of misdirection as a cognitive process that manipulates your beliefs about what you are experiencing. To do so effectively, misdirection exploits many of our mind's limitations. The key is that these limitations must be counterintuitive: once we become aware of them, we start attributing the effect to our limitations rather than to magic stuff.

HOW DOES MISDIRECTION WORK?

Misdirection lies at the center of magic theory, and magicians such as Arturo de Ascanio, Joe Bruno, Dariel Fitzkee, Al Leech, Jason Randal,

S. H. Sharpe, and Juan Tamariz have developed comprehensive theories about it.[41] But these frameworks are intended to help fellow magicians improve their performance and so are primarily concerned with ways to improve various tricks. In recent years, however, scientists have started to investigate misdirection by linking it to our understanding of the mind. This is my main focus here.

The first such theory was proposed by Peter Lamont and Richard Wiseman, who define misdirection as "that which directs the audience toward the effect and away from the method."[42] Their theory broadly distinguishes between two types of misdirection: *physical misdirection*, which deals with the control of attention, and *psychological misdirection*, which involves the remaining psychological processes.

When most people think of misdirection, they are thinking of the physical variety. This involves a magician manipulating your physical surroundings to create areas of high interest that will capture your attention and so prevent you from noticing things elsewhere. Lamont and Wiseman distinguish three types of physical misdirection, involving passive, active, and temporal diversions of attention. *Passive misdirection* uses anything that attracts attention in its own right—for example, a novel prop or a brightly colored object.

Active misdirection also controls our focus of attention but does so via the social interactions of the magician rather than the physical properties of the environment. For instance, the magician may use his eyes to direct your attention toward the areas where he is looking. *Eye gaze* is a very important and powerful misdirection tool, and much research has been carried out on how magicians use their eyes to misdirect you.[43] One of the most powerful ways of doing so is to simply ask you a question.[44] Have you ever noticed how difficult it is to ignore someone who is asking you something? Many street sellers exploit this attentional quirk by asking a simple question before launching into their sales pitch. Indeed, in one of my research studies, more than half the participants looked at the magician's face when asked a question, even though they were explicitly told to avoid being misdirected and to keep their eyes on his hands.[45] A whole range of other social cues, such as gestures and body language, can also be used to create similar areas of interest.

Just as your level of attention can vary over space, it can also vary over time. It is simply not possible for us to continuously attend to something, and magicians construct moments of interest to exploit these fluctuations in attention. This is *temporal misdirection*. For example, people are less likely to pay attention if they believe that the trick has not yet begun or is already over.[46] After a joke or a magical effect, there is also a momentary relaxation of attention during which you are less likely to notice the actual method being used.[47]

Richard Wiseman and Tamami Nakano recently investigated people's fluctuations in attention by measuring their eye blinks during a magic trick.[48] The results showed that people's eye blinks are highly synchronized and frequently occur immediately after an impossible feat. Interestingly, people's blink rates are significantly higher during those moments when the magician is trying to conceal his actions. Blinking is associated with relaxations of attention, and this study beautifully illustrates how magicians exploit these temporal fluctuations in attention.[49]

Psychological misdirection, on the other hand, has a strong emphasis on preventing detection of the method used in the trick by means other than attention. For example, magicians often need their actions to appear natural. If a magician wants to vanish a coin using the French Drop, any unnatural action will arouse suspicion and attract unwanted attention, resulting in the possible detection of the method. There are countless other techniques—such as the Ruse and the Magician's Conviction—that also fall within this broad category.

Lamont and Wiseman's taxonomy is a great improvement on earlier efforts because of its explicit links between magic theory and human cognition. However, because its purpose is simply to highlight misdirection principles for nonmagicians, it is somewhat lacking in scientific rigor, and some of the categories seem rather arbitrary. For example, the idea that you can separate misdirection principles into psychological and nonpsychological processes does not make much sense because attention and perception are no less psychological than memory and reasoning.

If we want to truly understand the psychological processes that underpin misdirection, we need to use existing models of the mind. Together with several colleagues, I have proposed a taxonomy of misdirection based on the kinds of processing that psychologists believe are used in human perception and cognition: when confronted with a magic trick, you first *perceive* the relevant sensory information, then *store* key aspects of it in your memory, and then perhaps use this to *reason* about how the trick was done.[50] A magician can prevent a spectator from discovering the method by simply manipulating any one of these processes.[51]

These categories define misdirection in terms of the psychological mechanisms affected (figure 2.3). The first set of principles manipulate your perception, preventing you from perceiving selected parts of the performance. Many kind of physical misdirection fall into this category. However, although attention clearly plays a very important role in determining what we perceive, many nonattentional psychological mechanisms do so as well (see figure 2.4). For example, *masking* is a technique whereby the magician prevents you from perceiving an event by physically obstructing your view. The magician might put his hand into his jacket pocket while turning to one side, which interrupts your line of sight. No matter how much you try to attend to the hidden hand, you simply won't be able to see it.

A physical barrier is not the only way to mask; other techniques can also prevent you from seeing an object or movement. For example, the Paddle Move is a technique in which the magician combines a rotating movement (i.e., rotating a paddle from one side to the other) with a larger tilting movement. This allows the magician to show two sides of a paddle as being empty even though something is actually printed on one side (see figure 2.5). Andreas Hergovich and his colleagues have conducted very carefully controlled experiments using three-dimensional animated graphics to show how the correct combination of rotation and tilting can entirely mask the rotation.[52]

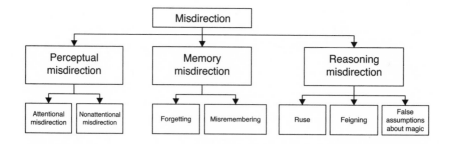

Figure 2.3

Taxonomy of misdirection overview (adapted from Kuhn et al., "Psychologically-Based Taxonomy")

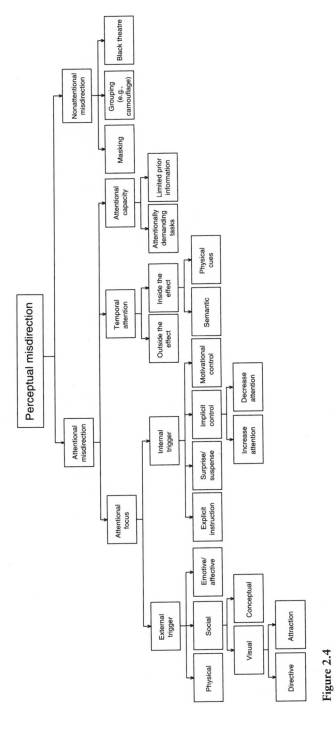

Figure 2.4

Perceptual misdirection (adapted from Kuhn et al., "Psychologically-Based Taxonomy")

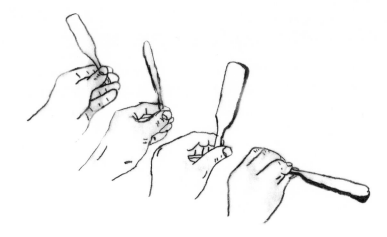

Figure 2.5

Paddle Move: As the hand rotates, the paddle rotates by 180 degrees. The large rotation masks the small rotation.

Masking is not limited to the visual domain: pickpockets often use tactile masks, such as pressing your wrist, to prevent you from noticing that they've stolen your watch. Other nonattentional principles include camouflage and perceptual illusion, which also prevent you from noticing the method but are unaffected by the amount of attention you give to them.

Dividing perceptual misdirection into attentional and nonattentional processes has important theoretical and practical implications. Although you can exert a substantial amount of control over what you attend to, you are much less likely to influence nonattentional processing, even by changing your focus of attention. For example, you simply won't be able to perceive an event that has been obscured by a physical barrier or perceptual mask, no matter how much you attend to it.

Attentional misdirection naturally plays a very important role, and even though it is very difficult to define "attention," three distinct aspects of attentional processing can be easily manipulated.[53] To begin with, magicians can manipulate the *focus* of your attention, which relates to *what* you attend to. As depicted in figure 2.4, there are countless ways in which this can be achieved. (We will discuss this in more detail in the next chapter.) Magicians can also manipulate *when* you pay attention and the *amount* of attention that you dedicate to seeing what is in front of you. For example, I might give you an attentionally demanding task, such as counting lots of different cards, which will prevent you from allocating attentional resources elsewhere and noticing the secret method.[54]

Even if you accurately perceive an event, however, there is no guarantee that you will accurately remember it later on. Our memories of an event are selective reconstructions, based on fragments of remembered experience rather than complete representations.[55] Our second category therefore involves memory, which includes any technique that manipulates *how* (and whether) you remember an event. For example, I can use particular techniques to prevent you from remembering certain details of an event or even make you remember things that never occurred. We will discuss these types of memory illusions much more in the next chapter; for now, I just want to point out that much of misdirection involves manipulating what you believe you remember of an event.

Even if you somehow—in spite of all odds—create an accurate memory of a magic trick, this still does not guarantee that you will discover its secret. Each member of the audience brings a set of preexisting assumptions about the nature of the world (including magic) to the performance. Although some of these assumptions will be correct, others will not. Manipulating these assumptions—and the reasoning based on them—forms our third category of misdirection.

For example, a common principle is to give you the impression that a trick involved very little planning. Thus, tricks are often designed to appear impromptu, preventing you from suspecting an elaborate setup of some kind. But magic tricks usually involve considerable preparation; nonmagicians typically fail to appreciate just how much work goes into creating them. This is of course intentional, because magicians don't want you to attribute the effect to an elaborate setup.

Teller has a very nice example that illustrates this point. According to him, people "will be fooled by a trick if it involves more time, money and practice than you (or any other sane onlooker) would be willing to invest."[56] Teller describes a trick in which five hundred live cockroaches were produced from a top hat during a performance on *Late Night with David Letterman*. The preparation for this apparently simple trick took weeks and cost vast amounts of money. For example, an entomologist was needed to provide slow-moving, camera-friendly cockroaches. A secret compartment also had to be created out of special foam, one of the few materials that cockroaches can't cling to. There also needed to be a sneaky way of smuggling this compartment into the hat. Most sane people would simply not go to such trouble, thus they struggle to imagine who would. This concept of putting unimaginable work into an effect is central to magic, and it is beautifully captured in Christopher Nolan's motion picture *The Prestige*, wherein two rival magicians nearly sacrifice their lives to create the perfect illusion.[57]

Teller's cockroach trick illustrates the amount of hard work often needed for a perfect trick. But can a magic trick be too perfect? This question, raised by magician Ricky Johnsson back in 1971, continues to cause endless debate among magicians.[58] The basic argument is that some tricks are simply too perfect to be true, and so people will easily discover the secret. If so, it then follows that you can create stronger effects by making the trick less perfect. At first sight, though, this theory seems odd. Why would anyone purposefully make their trick less perfect?

Cyril Thomas and I recently set up an experiment to test this theory using a simple mind-reading trick. Cyril asked a spectator to freely select a card, after which he explained that he would read her mind. After a few intense moments of staring into the woman's eyes, he revealed the name of the card. It was a flawlessly performed routine, and he discovered the card without any mistakes. Cyril is one of the most skillful card sharks I know, but rather than relying on sleight of hand, he used a very simple method: a deck of identical cards, ensuring that you will always chose the same card. Nearly 75 percent of the participants discovered the correct method, suggesting that it was pretty obvious. But here is the surprising part: when Cyril performed exactly the same trick but named a wrong card before revealing the correct card, only 30 percent of the participants discovered the correct method. These results are striking: making the trick less perfect considerably reduced the chance that people would notice the correct method. What is even more striking is that participants also found this version of the trick much more impressive than the perfect one. The too-perfect theory may be less crazy than it seems.

The Ruse is a less controversial way to misdirect reasoning. Here, the magician performs a mock action that provides a natural explanation for a visible action needed to carry out the method. For example, I might secretly palm a coin in my hand and would like to get rid of it. I could, of course, just put my hand into my pocket, but this would seem suspicious and attract attention. Rather than simply putting my hand into my pocket, I could use a ruse to justify why I am reaching for my pocket, such as pretending to pick up some magic dust. As my action is now justified, people will take less notice of it and won't remember that my hand ever went into my pocket.[59] Sander Van de Cruys and colleagues even argue that a ruse can change the way that we perceive an action.[60] For example, we can perceive an object as being in the foreground or the background but never as both. Van de Cruys suggests that the same principle applies to actions: an action can be perceived as a *putting* action or a *fetching* action but never both. Thus, once the audience interprets my hand as going into the pocket to fetch something, they simply cannot see this as a putting action.

There are countless other misdirection principles that involve manipulating your reasoning processes, and it is not my intention to examine them all here. But I would like to discuss one final principle, simply because it is my favorite: *dual reality*.[61] Imagine I'm about to display my incredible hypnotic powers. I ask a volunteer to help me out on stage and give him a quick hypnotic induction, telling him that in a few moments, he will lose his ability to read. I explain that whenever he looks at a piece of paper, his mind will simply go blank. I then write the word "chair" on a piece of paper, and although this word is clearly visible to the audience, if I turn the card to the hypnotized volunteer, his mind goes blank. No matter how hard he tries, all he sees is a blank piece of paper.

There are countless magic tricks in which the magician picks a volunteer from the audience, with the rest of the audience then observing the performance through the volunteer's eyes. In these types of tricks, there is always an implicit assumption that this volunteer experiences the same sequence of events as everyone else. For example, you assume that the hypnotized volunteer is looking at the piece of paper that you (and the rest of the audience) saw a few moments before. This is based on our natural assumption that all people must experience the same reality. But what if I secretly switch the piece of paper with a blank sheet? In this case, the hypnotized volunteer is really looking at a blank piece of paper, but you assume that the hypnosis (the pseudoexplanation) is causing his vision to go blurry. In reality, he cannot see the word because there is no word to be seen. You are both experiencing a very different reality, but the magician uses linguistic subtleties to convince you that you are both experiencing the same event.

I have described a relatively harmless trick, but fake spiritualist healers often use dual reality to create the illusion that they have real magical powers. For example, Derren Brown has discussed how an evangelical preacher may ask a partially blind person to join him on stage, while conveying to the congregation that the person is fully blind.[62] If you are fully blind, you won't be able to notice any difference between a dark room and a brightly lit one. However, partially blind individuals benefit from good illumination, and there are lots of natural ways in which you can improve their vision (without divine intervention). If the spiritual healer turns up the light, the partially blind man will indeed see more clearly, and when asked whether he can now see, his honest response will be yes. The partially blind man is not lying when he claims that his vision has been improved, but the congregation, which assumes that he was previously fully blind, has witnessed a true miracle.

Misdirection is the key psychological principle used to create the magical experience. Our psychologically based taxonomy of misdirection offers a starting point to help bridge the gap between magic and science and to help us identify the underlying psychological mechanisms. I am sure that as we learn more about the nature of misdirection, many aspects of this taxonomy will change. But this is not to be feared. Indeed, one of the purposes of our taxonomy is to help magicians and scientists keep making better sense of the various ways that misdirection can be achieved.

There are a few final points about misdirection that we also need to consider. Although we have used discrete categories as the basis of the taxonomy, mental processes are all interdependent. For example, your perception of an event influences what you remember, and your memories in turn guide your reasoning and perception. In addition, certain misdirection principles influence cognitive functions at multiple levels. For example, the theory of false solution can potentially influence what we attend to and thus perceive, as well as affecting our more abstract reasoning processes.

More generally, a good magic performance involves multiple layers of misdirection, and these must be complemented with charismatic performance and choreography. Our taxonomy of misdirection clearly does not capture these factors. Along with his colleagues, Wally Smith, a computer scientist, psychologist, and magician, has recently started to model the misdirection used in more complex magic tricks and has shown how computer logic can help understand what is going on.[63] One of the things that is becoming apparent is that combining different misdirection principles often creates a magical effect much stronger than the sum of its parts. In the final section, therefore, I would like to propose a new theory of misdirection, one based on the popular statistical inference method known as *Bayesian inference*. This theory allows us to combine different layers of misdirection and will hopefully go some way toward addressing complex tricks.

A BAYESIAN THEORY OF MISDIRECTION

As a psychology lecturer, I have learned that talking about statistics is the quickest way of losing someone's attention. However, before you go, I beg you to stick with me for a few minutes, because Bayesian inference could be the true secret behind misdirection.

In science and in our personal life, we often theorize about things, and we evaluate our theories using evidence. For example, you might be interested in whether it is possible to read another person's mind. Many scientists use Karl Popper's approach to evaluate this hypothesis: if the evidence does not match with the predictions of your hypothesis, that hypothesis must be rejected.[64] According to this approach, you reject your hypothesis regardless of how likely it was in the first place. Let us assume now that you have done your background research on telepathy and are pretty confident that it does not exist. However, to everyone's surprise, your study shows that mind reading is possible. You are now forced to conclude that telepathy actually does exist. By doing this, you have followed the traditional approach to hypothesis testing: you applied exactly the same logic regardless of your prior confidence in this hypothesis.

In recent years, an alternative approach, known as Bayesian inference, has gained popularity. I will not go into the details of this because the mathematics are horrendously complicated and of little concern to us here. The key point of Bayesian inference is that our prior beliefs in a hypothesis can have a direct impact on how we treat the evidence for or against it. According to Bayesian inference, you should discount a hypothesis that is extremely unlikely even if evidence supports it. Likewise, if you have a strong belief in a hypothesis beforehand, you should not reject it simply because it does not accord with the evidence. Bayesian inference can do this because it is about probabilities rather than black-and-white absolutes. Indeed, there is a very clever mathematical formula (known as Bayes's rule) that can inform us about the extent to which the given evidence should change the likelihood that the hypothesis is true. This means that if your study shows that telepathy is possible, you don't need to accept its existence. Instead, you merely adjust your certainty about it and consider it to be a little bit less impossible than before.

Think back to the illusion in which I pretended to use hypnosis to prevent a person from reading a word. This effect relies on your conviction that the hypnotized person is looking at a real word. To solidify the erroneous belief, I might add a small *convincer*: after the piece of paper has been switched for a blank piece, I might pretend to have forgotten the word and then quickly look at the blank paper as if to remind myself of the word. If you are feeling pretty confident that the piece of paper has the word written on it, this subtle convincer will enhance your belief. But the convincer is far less effective if you know that I've switched the paper. Simply looking at a blank piece of paper and pretending to read a word will not convince you, because the convincer's effectiveness depends on your prior beliefs. This process is very similar to Bayesian inference because the likelihood that you will change your theory depends not only on the evidence itself but on the prior confidence in the hypothesis or belief.

This Bayesian theory of misdirection is very much in its infancy, and many of its details still have to be worked out. However, I believe that it captures many of the key properties of misdirection, and because it is a computational theory, it's one that can be tested empirically. Most importantly, however, it illuminates parts of the complexity of misdirection in practice.

In chapter 1, we concluded that magic creates cognitive conflict between the things we believe we have experienced and the things we believe to be impossible. There are two ways in which this conflict can be handled. The first involves changing our beliefs about what we consider to be possible. In the next chapter, we will discover that many of our unconscious beliefs are much more magical than we think and that changing those beliefs is incredibly difficult. The second way to handle cognitive conflict is to change our experience: altering our perception about what caused the effect, such as by interpreting the event in terms of magic stuff. In this chapter, we discovered why we readily accept illusory causes and explored how magicians use misdirection to manipulate our causal reasoning so that we rule out any rational explanations that might come to mind, leaving only the magical ones. One of the main reasons why we are easily tricked into accepting these magical explanations is because we are genuinely deluded about our own cognitive abilities. In the next four chapters, we will explore some of these self-deceptions and see how magicians exploit them to create magic.

$\textcircled{3}$

THE BELIEF IN REAL MAGIC

UP UNTIL THE MID-NINETEENTH CENTURY, there were few individuals who could talk with the dead—at least, there were few who claimed that they could. But in 1848, a rather peculiar set of events dramatically changed our interactions with the spirit world. That year, two young sisters living in a small farmhouse in Hydesville, New York, discovered that they shared their room with a mysterious spirit.[1] The sisters, Margaret and Kate Fox, first became aware of its presence through strange nightly knocking. The spirit showed a clear fondness for the sisters: in their presence, the mysterious rapping began to occur even during the day, originating from the floor, walls, and even furniture.

To summon the spirit, Kate would knock several times on the floor; the spirit would then respond with the same number of raps. The sisters developed a form of Morse code that let them communicate with the spirit. The girls would ask it a question, after which the infallibly omniscient spirit would answer either yes or no, as indicated by the number of raps. This questioning eventually revealed that the spirit belonged to a man who had been murdered and was buried in their cellar.

The neighbors soon started dropping by to communicate with their deceased relatives, and news of the Fox sisters' abilities quickly spread through the small town. Their older sister, Leah, recognized the financial opportunity that her sisters' gift provided, and so with the help of a professional medium, the Fox sisters refined their skills. With the spirits no longer confined to the haunted farmhouse, the sisters organized public demonstrations that allowed paying customers to connect with the spirit world. People flocked to see the sisters, and the money followed. Of course, not everyone was convinced that their abilities were genuine. There were numerous investigations into the phenomenon, but any attempt at debunking the sisters only fueled publicity, spreading the news about them even further.

Because communing with the spirit world offered considerable financial gain, other mediums suddenly began to spring up throughout the United States. This fierce competition generated ever more sensational ways of communicating with the dead. Spirits now started responding in small sessions—paid, of course—that were known as séances. But the Fox sisters were not to be left behind in this battle of the spirits. They discovered ways to communicate by mysteriously moving the séance table and writing messages onto slate tablets. They even managed to cajole the spirits into playing musical instruments. These manifestations all took place under the cover of darkness, and they provided the believer with a "genuine" opportunity to communicate with their deceased loved ones. By the mid-1850s, spiritualism had become a mainstream movement, with many eminent men defending it in public.

In 1888, forty years after the Fox sisters' initial claims, Margaret admitted that their spiritualist effects were a fraud: the spirits' rappings were actually caused by cracking the joints in her toes and knees.[2] But although this revelation ended the Fox sisters' careers as mediums, it did not end people's belief in spiritualism. Believers became convinced that the retraction itself was a hoax, possibly due to less sociable spirits who wanted to lessen contact with their realm. And so the Fox sisters' legacy lives on. Even in today's world, spiritual readings and séances still take place, and a quick browse in your local New Age shop will verify that spiritualism is far from dead.

There is often a thin line separating what people consider to be real magic from what they designate as fake magic. We can only fully understand magic by understanding how some people can straddle this line— how they can genuinely believe in things they know to be impossible. In this chapter, we will explore the psychology underlying this phenomenon. Understanding magical beliefs certainly has implications for understanding magic itself, but perhaps more importantly, it provides rather surprising insights into our everyday behavior.

MAGICAL THINKING

The story of the Fox sisters illustrates that some people endorse beliefs in things that they know to be impossible. Put another way, they genuinely believe in magic. *Magical thinking* refers to thinking based on the belief that certain actions can influence objects or events, even though no cause connects them.[3] For example, scientists agree that we cannot communicate with the dead, meaning that beliefs in conversing with deceased people are therefore magical. Similarly, there is no evidence to support the existence of telepathy or remote viewing, and so beliefs in these are also considered magical. It is important to note that magical beliefs are not just beliefs that are scientifically wrong—they are scientifically impossible. For example, the mistaken belief that penguins can fly or that vitamin C can prevent you from catching the common cold are not inherently impossible, and we therefore don't consider them to be magical.

As with most definitions, defining magical belief is rather tricky. One problem is that our beliefs about the world continually change. Due to increased knowledge, we now have rational explanations for things that would have appeared magical just a few decades ago. Moreover, we all differ in what we believe to be possible. For our current discussion, then, I will define magical belief rather broadly as a belief in something that has been shown to be impossible on scientific grounds. I hope that scholars from other disciplines will forgive me for my rather sloppy definition. But as there is no universally accepted definition, I am inclined to just use my own.[4] Another interesting feature of many magical beliefs is that people often know that the phenomenon in which they believe is entirely impossible.

Let us start by briefly looking at one aspect of our lives that is heavily influenced by magical beliefs: religion. Much of religious experience (even of the modern kind) includes a sense of awe about the forces that cause seemingly inexplicable events in the world—similar to the sense of wonder created by a magic trick. And many religious scriptures describe miracles that seem much like magic tricks: the book of Exodus, for example, has a detailed description of Aaron turning the water of the Nile into blood and making a plague of frogs descend upon the land. The power by which religious figures are believed to act on the world seems a lot like the "magical stuff" discussed in chapter 2. Many of our thoughts and concepts about magic have come from studying religions and religious rituals.[5] However, for some strange reason, religions generally try to distance themselves from magic.

The classical scholarly view on magic sharply differentiates between magic and religion, and it considers the former to be a much more primitive institution.[6] This view assumes that although magical rituals and beliefs may have played an important role in so-called primitive societies, modern society has replaced such beliefs with causal thinking, resulting in modern religion and eventually modern science.

However, many people still believe in religion, including its more fantastical aspects, such as angels, gods, and other assorted beings. For example, 77 percent of Americans believe in angels.[7] There are many reasons for these beliefs, and although a full discussion of these is beyond the scope of this book, it is nonetheless clear that magical beliefs are still alive and well in our society. As an atheist, I have always struggled to understand how people can simultaneously endorse religious and scientific beliefs. But as we will discover in this chapter, all these beliefs can happily cohabit inside a single brain.

DO YOU HAVE TO BE MENTALLY UNWELL TO BELIEVE IN MAGIC?

Researchers often consider magical thinking to result from people not thinking about things correctly, either because of cognitive problems or because they are simply too young to know better.[8] In fact, many of our contemporary ideas about magical beliefs originate from the study of clinical disorders. For example, magical beliefs are related to the intrusive thoughts observed in people with obsessive-compulsive disorder (OCD). Individuals with OCD engage in repetitive behaviors, such as handwashing, checking that the stove is turned off, or touching a door handle a specific number of times before entering a room. Such behaviors are a form of magical thinking, as there is no reason for people with OCD to connect their behavior to the misfortune they are trying to ward off. A person can fully understand that repeated checking is not necessary yet still feel compelled to do so before leaving the house.

There may also be a link with schizophrenia, which involves a wide range of symptoms, including delusional thoughts, hallucinations, disorganized speech, and extreme social withdrawal. It often includes the belief that one's thoughts are being broadcast or that one can read others' minds. Psychologists typically consider these beliefs to be magical because they contradict our understanding of the world. But delusional thoughts may simply be mistaken beliefs and so would not necessarily be magical. On the other hand, schizophrenia has been linked to a psychological variable known as schizotypy, which in turn correlates with belief in the paranormal (i.e., phenomena such as telekinesis and clairvoyance that are beyond the scope of normal scientific understanding).[9]

Based on the above, it would be tempting to argue that those who endorse magical beliefs must suffer from some form of psychopathology. But although links do exist between magical thinking and some psychopathological symptoms, these links are generally rather weak.[10] Another thing to keep in mind is that magical beliefs are extremely common in the normal population. For example, a 2005 survey found that three-quarters of the American population endorsed at least one paranormal belief: 31 percent of respondents believed in telepathy, 32 percent in ghosts, and 41 percent in extrasensory perception (ESP).[11] If a connection between such beliefs and psychopathology were true, it would imply that most people are mentally ill.

Of course, you don't have to endorse paranormal beliefs to believe in the impossible. Many of our superstitions, such as the belief that walking under a ladder can bring bad luck, are common magical beliefs in the general population. For example, more than half of respondents in a 1984 poll conducted in London admitted that they avoided walking under ladders and touched wood for good luck.[12] Similarly, a survey of college students found that nearly one-third regularly engaged in exam-related superstitious behavior.[13] Superstitious rituals are also well documented and particularly prevalent in sports.[14]

Over the past five years, my colleagues and I have carried out research on magical beliefs among undergraduate students, and I have been truly amazed—or maybe I should say dismayed—by the results. We were inspired by Victor Benassi and his colleagues, who used magic tricks to study magical thinking back in the 1980s.[15] In collaboration with Christine Mohr and her PhD student, Lise Lesaffre, we have used magic tricks to create anomalous experiences and then asked observers to report on their experience.[16] These experiments, which are based on a fake demonstration of psychic abilities, have been some of the most fun and intriguing that I have ever done.

I told the students that they would be part of an experiment investigating a psychic medium.[17] I explained that the Anomalistic Research Unit at Goldsmiths, University of London, had a long tradition of investigating spiritualist and other paranormal claims. Although most of these claims had not held up under close scientific scrutiny, I said, we had recently found an individual who, although not perfect, had a hit rate significantly above chance. At this point, I introduced them to my friend Lee Hathaway, a professional magician with whom I have performed for many years.

On this occasion, Lee transformed from a polished magician into a campy soothsayer, using his conjuring skills to create a convincing séance. He used gimmicks to trick participants into believing that he could read the color of their aura. He also used cold reading to give people the impression that he was reading their soul, and he finished the routine by contacting a dead relative of a randomly selected volunteer. This last part was entirely staged; the volunteer was actually one of our stooges.

Describing this demonstration in words does not do it justice. Lee and I are both experienced performers, but we never imagined that this demonstration would have such a powerful effect. Whenever I perform magic, my audience knows that what they are seeing is not real. But this time, I used my scientific authority to deceive students into accepting magic tricks as real. The first time we ran the experiment, we were both rather shaken. Most of our participants genuinely believed that Lee possessed real psychic powers! We were particularly unsettled by just how easy it had been to manipulate highly educated people. When you run experiments on human subjects, it is important to debrief people thoroughly once the experiment is over. Most of our participants were genuinely surprised and somewhat disappointed when we told them that they had been tricked.

We used a range of tests to measure participants' beliefs, and we also asked them to write down their thoughts about the experience. The majority of our participants believed that Lee's psychic abilities were genuine. It is important to note that all of our participants were undergraduates at a prestigious UK university, and yet over half of them were willing to believe something that was impossible. These results vividly exemplify that we cannot simply attribute magical thinking to cognitive deficits.

One of our most surprising findings was that people were happy to endorse magical beliefs even when we told them that Lee was a fake.[18] Prior to seeing the demonstration, half of our participants were informed that Lee was a magician who used tricks to pretend that he was a psychic. To our absolute astonishment, this information had virtually no effect on their beliefs. Some of our participants, who just moments before had copied out instructions telling them that Lee was a fraud, were fully convinced that he was a genuine psychic. We have since run several similar experiments, with similar results. It was only when we told people how the tricks had been done that they started questioning the psychic's powers.[19]

It is important to note that not all of our participants were convinced by Lee's powers. There was a strong correlation between people's preexisting beliefs and the extent to which they attributed the demonstration to Lee's genuine psychic abilities. People who believed in spiritual phenomena were more likely to interpret the anomalous event as genuine, regardless of what they had been told beforehand.

Why do people believe in spiritual phenomena, even though these are scientifically impossible? Part of the answer may involve memory distortions that render the phenomena consistent with prior beliefs. Back in the 1880s, Richard Hodgson and S. J. Davey showed that people often struggled to remember crucial details of a séance.[20] They did this by holding fake séances and afterward asking the sitters what had occurred. Hodgson and Davey reported that many sitters omitted important details about the event and even reported witnessing genuine paranormal phenomena.[21]

More recently, Richard Wiseman and his colleagues also showed that believers in the paranormal are more likely to misremember paranormal events.[22] They set up a fake séance and used trickery to create various effects (e.g., a hidden assistant moved objects with a long stick). The sitters were later asked about what they had witnessed. The results showed that believers in the paranormal were more likely to have memories of events consistent with their beliefs. For example, they were more likely to claim that the handbell had moved, even though it had not. A later study similarly found that believers in the paranormal were more likely to interpret and remember ambiguous events in ways that confirmed their preexisting paranormal beliefs.[23] But such memory biases are not the whole story.

To show this, my colleagues and I recently carried out our own experiment to measure people's memories of spiritual performance.[24] We asked our volunteers to judge the accuracy of different types of readings, and we evaluated the extent to which they felt that the phenomena could have occurred by chance. We expected that believers in the paranormal would be more likely to misremember inaccurate information that supported their beliefs. Curiously, though, we found no difference between skeptics and believers; both groups were surprisingly good at remembering what they had seen. Why might this be? Our demonstration used more sophisticated deception and so created a more anomalous experience than those of previous studies. It is therefore possible that this was sufficiently powerful to elicit paranormal beliefs in most of the individuals, not just the believers. It is also likely that people's momentary beliefs about the experience will influence how they remember the event in the future, for as we will see in subsequent chapters, our memories of events are much more malleable than we think.

The traditional view of cognition assumes that magical beliefs are restricted to a small group of individuals. But as has been shown here, this assumption does not stand up to scrutiny. Paranormal and superstitious beliefs, which are both magical, are held by large segments of our society. By using trickery and deception, we can even manipulate large numbers of highly educated people into endorsing magical beliefs. Moreover, as we shall now see, magical beliefs exert a surprisingly large influence on our everyday lives.

HOW MAGICAL BELIEFS INFLUENCE OUR LIVES

Anthropologists, historians, and sociologists have played an important role in developing our ideas about magic.[25] For example, at the beginning of the twentieth century, Sir James Frazer and Marcel Mauss wrote influential books about magical practices and rituals in cultures across the globe.[26] They also proposed laws—known as *sympathetic magic*—to explain many of these practices. One of these was the *law of contagion*, which states that things that have been in contact may influence one another via the transfer of a magical *essence*. This influence can remain after the physical contact has been broken and may even be permanent. Importantly, this form of magical contagion applies to situations in which there is no scientifically plausible reason for one entity to have influenced the other. For example, the Indian caste system asserts that food touched by anyone from the *dalit* caste is contaminated (hence the term "untouchable"), even when there is no possible way any contamination could have happened. The law of contagion can also manifest itself in more positive ways. In many religions, for instance, being physically touched by a holy figure is thought to transfer a positive essence, such as a blessing.

It is tempting to believe that these forms of magic are only found in non-Western cultures, but the law of contagion plays an important role in our own society. Many of us use alternative medicine even though the mechanisms by which it is claimed to work conflict with our scientific understanding of the world. For example, in 2007, the National Health Interview Survey estimated that approximately $2.9 billion is spent on homeopathic medicine in the United States every year.[27] Such medicine involves the repeated dilution of a solution that contains an "active" ingredient, resulting in tinctures that are barely distinguishable from water. Indeed, it is believed that the greater the dilution, the more potent the medicine. Interestingly, homeopathic medicine uses two principles to account for this, both with similarities to sympathetic magic. The first relies on the belief of "like cures like," whereby a substance that causes a symptom can cure the same symptom when taken in a more diluted form.[28] The second principle relies on a form of contagion in which the substance's essence is transmitted from one solution to the next by contact.

There is no scientifically plausible mechanism that can explain why these principles might work. Indeed, the suggestion that a substance diluted to such extreme levels could still affect us goes against much of current science.[29] Yet there are countless studies showing positive homeopathic effects for a wide range of symptoms.[30] How can this be? An interesting possibility is that homeopathic medicine works not because of its magical tinctures but because of our belief in those tinctures. Large-scale meta-analysis has shown that a homeopathic tincture has exactly the same healing power as a placebo, suggesting that its effects are all in our mind.[31] Simply believing that you can be cured by an impossible or magical process can sometimes turn out to be rather effective.

I am deeply skeptical about the rationale underlying most forms of alternative medicine. When suffering from a headache, I always reach for paracetamol rather than a homeopathic tincture. More generally, I am often cynical about all principles that have not been scientifically proven. I like to think of myself as a rational person, immune to magical beliefs, but I am deluding myself. I am just as susceptible to the placebo effect as anyone else, and many of my own thought processes rest on deep-rooted magical beliefs. Let me give a quick example that many readers can likely relate to: I have been happily married for nearly nine years. Like most married men, I am strongly attached to my wedding ring; losing it would be devastating. Why am I so attached to my ring? There is nothing particularly special about it, and I could easily replace it with an identical one. Yet somehow, the replacement simply would not feel the same. Could it be that I am attached to some form of metaphysical magical essence?

Most of us have objects that we are attached to, but in most instances, what we actually treasure is the nonphysical essence that we associate with these objects. This magical attachment develops at an early age. Most children have a special teddy bear or blanket that they use as a comforter; losing such an object can be very distressing. Bruce Hood and Paul Bloom have carried out fascinating experiments to investigate this phenomenon.[32] They invited parents and children (aged three to six years) to their lab and asked them to bring along several toys, some of which the children were attached to and some of which they were not. The researchers then demonstrated a magical copying machine that let them duplicate different kinds of items. For example, they placed a toy rabbit into a box, closed the lid, and pressed some buttons. When they opened it, an additional rabbit appeared inside. Although the magic box would not have won a prize at any magic competition, all of the children were convinced of its effectiveness.

The children were invited to place their own item into the box and were then asked whether they would like to take home the original or the replica. For a nonattachment item, 62 percent chose the duplicate. But for an attachment item, such as their comforter, only 23 percent did so. In fact, 20 percent of the children refused to place their beloved item into the box at all, for fear of it being duplicated. These results clearly show that children value the original object much more highly than a physically identical replica.

In a subsequent experiment, the researchers duplicated a precious metal spoon and asked the children whether they preferred the original or the duplicate. Here, the spoon contained no magical essence, and 82 percent of the children valued the items equally, demonstrating that they believed the copy to be a true replica. But in another scenario, the children were told that the original item had belonged to Queen Elizabeth II. They were now over three times more likely to choose the original over the copy. Mere ownership by the queen apparently lent the metal additional value—a value that was magically tied to a particular object.

Adults are far from immune to these forms of contagion. Such magical thinking reveals itself in various ways, such as our irrational attachment to particular objects and our obsession with products that have been endorsed or touched by celebrities. A few minutes surfing the internet reveals just how much people are prepared to pay for magical essences that have transmitted by contagion: A fan paid $394,000 for the chair that J. K. Rowling sat on while writing the first two books in the *Harry Potter* series.[33] Scarlett Johansson's used tissue sold on eBay for £3,320.[34] These may be extreme cases, but the same principle affects us all.

Paul Rozin and his colleagues have carried out numerous studies showing just how susceptible we are to this form of sympathetic magic. For example, normal undergraduate students valued a T-shirt that had been worn by a loved one much more highly than one that had been worn by a less desirable person, which illustrates that this form of magical thinking is the norm rather than the exception.[35]

The principle of negative contagion is even more powerful than positive contagion, and there are some wonderful experiments illustrating the mechanism at work. Imagine sitting at a table with a nice cold drink in front of you, and your friend suddenly dips a sterilized cockroach into your beverage. I consider myself to be very skeptical about magic and am fully aware that a sterilized insect cannot contaminate my drink, yet the sight of the cockroach would certainly put me off my beverage. Paul Rozin and his colleagues showed that I am not alone.[36] Likewise, students preferred a drink containing the label "sucrose" over one with the label "cyanide," even though they were aware that the labels were applied randomly after the drinks had been poured.[37] Such results clearly show that even though we like seeing ourselves as rational individuals, our behaviors are instead deeply rooted in magical thinking.

A second law of sympathetic magic, the *law of similarity*, states that things that resemble one another physically share fundamental properties. This law assumes that images of an object, person, or animal are somehow equal to the object itself. For example, Frazer describes tribes who avoid eating the flesh of slow animals for fear of becoming slow and people in northern India who believe that eating the eyeball of an owl will allow them to see in the dark.[38] Voodoo magic involves both laws of similarity and contagion. Here, parts of your enemy, such as a hair or fingernail, are incorporated into a voodoo doll, and the process of contagion transfers your enemy's essence into the figure. The physical similarity between the voodoo doll and the person's body further ensures that any harm done to the doll will magically result in physical pain experienced by the person.

Few of us consciously admit to believing in the law of similarity, yet research has shown that this principle influences much of our behavior. If you are not convinced, ask yourself this question: would you rather eat a piece of chocolate shaped like a muffin or one shaped like dog feces? I know which one I would go for, and indeed Paul Rozin and his colleagues have shown that feces-shaped chocolate is overwhelmingly rated as less pleasant than muffin-shaped chocolate.[39] Similarly, people do not enjoy holding a piece of imitation vomit, even if they know that it is entirely made of clean rubber.[40] Although we rationally know that there is no relationship between the perceptual similarity of an object and its other physical properties, we simply cannot inhibit the emotional reactions that these objects elicit.

There is much evidence to support this idea. In the same paper, Rozin and his colleagues show that people often treat pictures of objects as if they were the real thing.[41] For example, students were required to throw darts at a picture of a person they liked (John F. Kennedy) or one they disliked (Adolf Hitler), and the researchers measured the accuracy of throwing the darts at the picture. The darts strayed about eleven millimeters farther from the bull's-eye when participants targeted JFK compared to Hitler. Similarly, people are rather hesitant to throw a dart at a picture of a baby's face.[42]

But surely people do not mistake a picture for reality? Although I am perfectly capable of recognizing the difference between pictures and reality, there are times when this boundary becomes more blurred. For example, while watching a horror movie, you become scared even though you consciously know that you are merely seeing a movie. Similarly, people are generally reluctant to tear up photographs of people they are close to, and I even feel bad about deleting pictures of my kids from my phone.[43]

Bruce Hood and his colleagues have conducted research that nicely illustrates this blurred distinction between pictures and reality.[44] They asked people to bring sentimental childhood items to their lab and then took a photograph of these items. They then blurred each photo so that it could no longer be recognized by anyone who did not already know what the object was. Participants were then asked to cut these pictures in half while their anxiety level was measured. Results showed high levels of anxiety for pictures of sentimental objects but not for pictures of other familiar objects. This effect was even found when participants were not being observed, suggesting that their anxiety was not due to a public display of destruction. Hood and his colleagues suggest that these findings support the notion that although most adults explicitly reject the principles of sympathetic magic, such principles still remain in the adult mind and can occasionally influence our behavior and emotions.

So far, we have seen that magical beliefs play a much more dominant role in our daily lives than we think they do. Let us now look at the origins of these magical beliefs and the reasons why we are so reluctant to abandon them.

MAGICAL THINKING IN CHILDREN

I regularly perform magic tricks for my own children, and I am fascinated by how their reactions have changed over the years. For example, at the age of five, both Ella and Joe were pretty convinced that I could do real magic. This belief was quite apparent on Ella's fifth birthday. She was excited when she unwrapped her present, which contained a magic wand. But she was adamant that I would have to put some magic into the wand before she could use it properly. Three years later, Ella is much more skeptical about my magic powers. For example, as I demonstrate my mind-reading powers at the breakfast table or make the odd piece of toast disappear, she is pretty sure that I am simply tricking her. But she has not yet abandoned all her magical beliefs; she still believes in Santa Claus, the Tooth Fairy, and even the Easter Bunny. My youngest daughter, Mae, on the other hand, has just turned two, and even though she enjoys seeing balls disappear, I doubt she has a clear understanding of magic. Let us now look more closely at how magical beliefs change as children grow up.

Magical thinking plays an important role in Jean Piaget's developmental theories. He argues that magical beliefs are typically found in younger children, but as they learn more about the true causal relationships in their environment, children replace these magical beliefs with a scientific thinking process.[45] For example, Piaget observes that younger children often attribute consciousness to inanimate objects, such as a string that wants to unwind because it is twisted. Similarly, Piaget also observes that younger children often believe that desires can magically influence reality, such as when a young boy believes that "birds and butterflies in his father's illustrated manuals would come to life and fly out of the book.[46] Piaget's view on child development has been extremely influential. His observation that children's magical beliefs change as they get older is correct, but he misses several important aspects of human cognition. As we have seen in this chapter, adult cognition is strongly influenced by magical beliefs, showing that our belief systems are not entirely replaced by scientific thought processes. Let us now look at these magical beliefs in more detail, as well as the reasons why they develop in the first place.

Once you observe your children's everyday activities more closely, it becomes more obvious as to why they believe in magic. Most parents, including myself, actively encourage their children to believe in magical figures by reading them stories that contain fairies and other supernatural beings.[47] I go to great lengths to foster these magical beliefs and provide my children with false evidence to enhance them. For example, on Christmas Eve, I dress up in a Santa Claus outfit just to cover my tracks in case one of my kids wakes up as I sneak into their room. It is not only parents who encourage these magical beliefs, as most children's television programs are packed with magic.

Given that adults continually deceive their children, it is little surprise that most preschoolers believe in fictional magical characters. These magical beliefs generally peak when children are five to six years old, after which they gradually decline.[48] Experiments have shown that around that age, children can be easily convinced into believing in a new fictional character. Jacqueline Woolley and her colleagues invented a magical character called the Candy Witch, a nice witch who visits children's houses on Halloween night and replaces the candy that the children have collected with a new toy.[49] Teachers and parents helped propagate this myth, and several parents replaced the candy with a toy at night. Within about two weeks, most of the children believed in this magical creature, and their belief in the new Candy Witch rivaled better-established figures such as Santa Claus and the Easter Bunny.

I sometime question my older daughter's belief in such mystical creations and wonder if it might be driven by alternative motivations. Most of the fantastical figures in which my children believe bring some sort of gift (e.g., presents, chocolate, or coins), so it is conceivable that their magical beliefs are driven by these positive rewards. In a clever manipulation, Woolley and her colleagues measured children's excitement about getting a new toy versus their disappointment about losing candy to the Candy Witch. Their analysis revealed that among younger children (three to four years old), belief in the Candy Witch was independent of their preference for a toy versus candy. However, for the older children (four to five years old), the expected reward (gaining the toy) and the expected consequence (losing the candy) correlated with their belief in the Candy Witch. These results clearly show that as children get older, they become more strategic in the beliefs that they endorse.

As children grow older, they develop a greater understanding of the real world and the natural physical explanation of events. For example, they learn that objects don't simply move or disappear on their own. At the same time, they also become aware of conjuring tricks that can be learned by anyone.[50] For example, I have noticed that when I use sleight of hand to vanish a ball and make it reappear for my youngest daughter, Mae, she will grab the ball and blow on it in the hope that it will disappear. My older children, on the other hand, will take the ball and use some form of sleight of hand and misdirection to create the illusion that the ball is disappearing. Unlike Mae, they clearly have some understanding of trickery.

Even though children have some rudimentary understanding of magic tricks, children's magical beliefs are very strong. For example, I have fond memories of receiving a Superman outfit for my sixth birthday. Despite the fact that I was fully aware that the cape would not allow me to fly, there was still some doubt at the back of my mind. Wearing my cape and the rest of the suit, I climbed up my bunk bed ladder and took a step into the unknown. Even though I knew that the cape was unlikely to work, I was still rather disappointed when I immediately plummeted to the ground. At the time, I would have told everyone that I did not believe in magic. Yet the magical thoughts were still present.

Eugene Subbotsky has conducted some preliminary research showing just how strong the belief in magic is at that age.[51] In his studies, children aged four to six years were told a story about a girl who had a box that could turn pictures into real objects. When questioned, most children denied that this was possible. But when the experimenter left the room, 90 percent of them tried to turn pictures into objects and were bitterly disappointed when it did not work. In a follow-up study, the researchers placed a plastic lion on the table and explained that the magic table could turn toy figures into real ones. When directly questioned, most children above the age of four claimed that this was not possible. The experimenters then used a concealed device to animate the lion, after which most of the children either ran away (fearing that the lion was coming alive) or used a magic wand that they had been given to stop the lion from moving. Although older children often explicitly deny that they believe in magic, their behavior suggests otherwise.

As children grow older, they give the impression that they no longer believe in magic. But it does not take much to change their beliefs. Subbotsky has shown this nicely in a series of experiments using a magic box.[52] The researchers placed a stamp inside this box, after which the magician cast a spell ordering the stamp to be burned. When the box was opened, the children found a half-burned stamp. Before seeing this trick, most of the children claimed that this type of magic could not happen in real life. But after seeing the trick, most of the five-to-six-year-olds abandoned their skeptical view and acknowledged that this was real magic. The nine-year-olds were more cynical, and only half of them acknowledged that the trick had been created through real magic. After the trick was exposed, the nine-year-olds quickly recovered their initial skepticism. However, only half of the five-year-olds accepted the nonmagical explanation, with the others continuing to believe in magic. Subbotsky suggests that when questioned, children show rational and logical thinking because this is what is expected of them. However, this disbelief in magic is only superficial; as these experiments demonstrate, children are easily persuaded that magic is real.

Thus far, we have seen that adults often deny believing in magic, but on closer inspection, much of our behavior is more magical than we think. Subbotsky suggests that in adults, magical beliefs are simply suppressed and can be reactivated given the appropriate conditions.[53] He also suggests that when denial of a magical belief is costly, adults are happy to give up their belief in the power of physical causality and view the world in terms of magical explanations.[54] More importantly though, these results clearly show that magical and scientific beliefs can happily coexist inside our minds.

WHY DO WE BELIEVE IN MAGIC?

Now that we have established that magical thinking is deeply ingrained in our day-to-day thoughts and behaviors, let us take a look at why such thinking might exist in the first place. A full discussion of why people believe in magic is beyond the scope of this book, but I would like to briefly discuss some of the most prominent ideas on this topic.

Several people have suggested that magical beliefs offer an adaptive strategy for dealing with the complexity of our everyday lives. Most aspects of our lives are driven by science and technology, and as we grow up, less and less becomes possible. Teenage years represent one of the most exciting periods of our lives, a time with seemingly endless possibilities.

As we become adults, we get bogged down by work and other responsibilities, and for many of us, magic offers a chance to escape the mundaneness of everyday life. Magic pushes the boundaries of what is possible, and many find the idea of entering a world that is less constrained by the laws of reality to be a very appealing proposition. We are strongly drawn to magic in film and literature because entering a magical world offers a very interesting alternative reality, which many of us find comforting.[55]

In children, magical beliefs provide fuel for imaginary role playing and fantasizing that helps them to cope with the chaos of their subconscious desires and to master difficult problems.[56] This helps children maintain a feeling of independence and power, and similar concepts also play a role in our adult lives. Magical beliefs can help us adults deal with complex situations that we would otherwise simply fail to understand, and they can make the inanimate world more understandable and humane.[57] For example, human-computer interactions rely on a deep-rooted magical belief that is typically known as the *user illusion*.[58] Every time you empty your computer's trash folder, you happily accept the magical belief that the files within have been deleted. Accepting this magical user illusion is far more manageable than having to deal with the complexity of computer programming.

Another aspect is the illusory sense of control that magic provides, with magical beliefs offering a helping hand in situations beyond our rational control. Control is an important coping strategy, and a lack of control can lead to mental health issues such as depression. Bronisław Malinowski argued that magical beliefs and superstitious behaviors allow people to reduce the tension created by uncertainty and help fill the void of the unknown. Malinowski noticed that the behavior of fishermen in the Trobriand Islands changed depending on where they fished. In the inner lagoon, fishing was straightforward, with little ritual. When fishermen set sail for the open sea, however, there were much higher levels of superstitious behavior, often involving elaborate rituals. The water in the inner lagoon was always calm and the fishing consistent, with little risk and, consequently, a high level of perceived control. Fishing in the open sea, on the other hand, was more dangerous, with prospects that were much less certain, resulting in a lower sense of control.[59]

More recent studies have provided further support for this connection. During the 1990–1991 Gulf War, researchers observed more magical thinking and superstitious behavior in people who lived in areas under direct threat of a missile attack, compared to those in low-risk areas.[60] In their study of superstitious rituals employed during high-stress examinations, Jeffrey Rudski and Ashleigh Edwards observe that the frequency of students' exam-related magical rituals increases as the stakes increase.[61] Intriguingly, students report that they frequently use these rituals while denying any causal effectiveness. Superstitious behavior therefore seems to give us the *illusion of control*, which can reduce anxiety during stressful situations and consequently improve performance. As with homeopathic medicine, many of these rituals might actually work, albeit through unintended or indirect mechanisms.

Few doubt that magical beliefs can provide an illusory sense of control, but why do normal people also develop and maintain magical beliefs in ordinary, nonstressful contexts? Jane Risen suggests that magical beliefs result from some of the shortcuts and heuristics that our minds use to reason about the world.[62] According to her, there is nothing intrinsically special about magical beliefs; they simply reflect some of the biases and quirks found in our everyday cognition. Let us now examine this theory in a bit more detail.

In recent years, psychologists have proposed that we use two fundamentally different mental processes to solve cognitive tasks. In his influential book *Thinking, Fast and Slow*, Nobel laureate Daniel Kahneman proposes that our reasoning and decision making rely on two separate mental processes. One of them, System 1, operates quickly and requires little cognitive effort. Rather than analyzing a problem in all its detail, it uses simple heuristics to come up with quick, intuitive answers. In many situations, this is an effective and reliable strategy. But as with any shortcut, it can lead to errors.[63]

An example of this is the *availability heuristic*, a cognitive shortcut that that helps us evaluate the importance or prevalence of an event based on the ease with which we can remember the appropriate information. Information that comes to mind more easily is weighted more heavily. This is why, for example, most people vastly overestimate the likelihood of dying from a shark attack. Such attacks are extremely rare; you are far more likely to be killed by a cow.[64] Yet unlike cow attacks, shark-related deaths are widely reported in the press and so pop into your mind more easily, thereby influencing your beliefs.

Although System 1 is fast, it is not necessarily accurate, whereas accuracy is much better with System 2, the other mental process. But System 2 operates in a controlled, step-by-step manner, making it rather slow and effortful. According to Kahneman, most of our day-to-day decisions are made through System 1, with System 2 intervening to override these intuitive assessments when they go wrong.[65] Unfortunately, however, System 2 is often too effortful, so that many of these wrong answers go unnoticed, especially when they seem like they're correct.

Let me illustrate this using a famous problem-solving task.[66] Try to solve the following problem: The combined cost of a bat and a ball is $1.10. The bat costs $1.00 more than the ball. How much does the ball cost? Before reading on, take a few moments to solve the problem. (No, really. Give it a try.) The answer that immediately springs to most people's minds is $0.10. But this is incorrect. If the ball cost $0.10 and the bat cost $1.00 more, the total would be $1.20, not $1.10. The correct answer is actually $0.05. Even though solving this problem does not require sophisticated mathematics, more than half of the participants at elite universities and more than 80 percent of participants at less selective universities answered it incorrectly.[67]

If you came up with $0.10 as the answer, you relied on System 1 and did not invest enough cognitive energy to check your answer. Had you done so, you would certainly have spotted the error because the problem is not particularly challenging. The fact that most people fail to check their answer suggests that System 2 is often lazy and inattentive. As we will learn throughout this book, there is huge pressure for the brain to save its cognitive resources. System 1 requires less effort and is much more likely to be used, even though it occasionally makes mistakes. People who come up with $0.10 as the answer have replaced "the bat costs $1.00 more than the ball" with a simpler statement: "the bat costs $1.00." According to Kahneman, most of our cognitive reasoning is carried out by System 1, but once System 2 spots a mistake, it corrects it and enables us to come up with the correct answer.

Jane Risen recently suggested that, in many situations, System 2 notices the mistake but still does not correct it, acquiescing to the erroneous conclusion.[68] The idea that you would continue to believe something that you know to be wrong sounds rather odd, but of course, this is exactly what we observe during magical thought processes. For example, when participants refuse to drink a beverage labeled "cyanide," they know that they are making an error, just as they know that cutting up a picture of a loved one causes no real harm. It is clear from participants' verbal reports that people realize that their feelings toward these objects are unfounded but that they feel them anyway. For example, I know that there is nothing special about my particular wedding ring, but I feel strongly about it nonetheless.

Jane Risen argues that superstitions and other powerful intuitions can be so compelling that we simply cannot shake them off, despite knowing that they are wrong. According to her, System 2 is not simply lazy and inattentive, it is also "a bit of a pushover"; it will not override the result of System 1 if the feelings associated with that result are too strong.[69] Many of the magical beliefs discussed so far, such as the law of contagion, occur because we rely on System 1's simple heuristics and employ them in situations where these rules do not apply. Even though System 2 knows they are wrong, it fails to correct the erroneous logic and thus acquiesces to magical beliefs.

The idea that you would believe something that you know to be impossible seems rather counterintuitive. However, this is only one of many strange and counterintuitive properties of our mind. It is important to note that Risen's new model of cognition does not apply exclusively to magical thinking and can explain a wide range of rather irrational behaviors.

For example, in 2015, British gamblers lost a staggering £12.6 billion.[70] In 2016, American gamblers lost even more: $116.9 billion.[71] People's probability judgments clearly have some rather irrational characteristics. Many of these judgments are based on System 1 responses, which people often know are wrong.[72] Imagine that you can win a prize by selecting a red marble from a bowl that contains both red and white ones, and you can choose whether you'd like to pick from a small bowl or a large one. The small bowl has ten marbles, one of which is red (a 10 percent chance of winning). The large bowl has one hundred marbles with fewer than ten that are red (a less than 10 percent chance of winning). You know the odds, which are clearly marked on each bowl. So which bowl would you choose? Rather surprisingly, over 80 percent of people chose the large bowl, even though they knew that the odds of winning would be lower.[73] We are evidently compelled to choose this bowl because it contains the larger number of winners. This is one of numerous situations in which System 1 makes a decision based on a heuristic (i.e., choose the situation with largest number of winners), while System 2, which knows the odds, fails to override this intuitive yet suboptimal decision. Likewise, sports gamblers are reluctant to bet against the favorite, even if the potential winnings of the underdog are higher.[74] Again, System 2's failure to override such decisions contributes to the astronomical profits made by casinos and bookmakers and influences consumer behavior and stock markets around the world.[75]

In this chapter, we have explored our beliefs in "real" magic and the important role these play in much of our day-to-day behavior. The current research on magical thinking challenges many traditional views of cognition—in particular, the view that childhood magical beliefs are replaced by rational and scientific reasoning in adulthood. Instead, it has become apparent that rational and magical thoughts cohabit deep inside our minds. Most previous models of cognition have struggled to accommodate the coexistence of magical and scientific thought processes, hence the need to revise our models of cognition.[76]

Understanding our magical beliefs also helps us understand the experience of performance magic, because witnessing a magic performance results in a coexistence of contradictory beliefs. As previously discussed, Teller describes magic tricks in terms of experiencing things as real and unreal at the same time, while Jason Leddington suggests that the experience of magic results in a conflict between our beliefs about the world and the automatic alief that the trick itself elicits. These ideas share many similarities with the theories of magical thinking discussed in this chapter. In light of this new research, the idea of simultaneously holding contradictory beliefs or experiences seems entirely plausible. It is tempting to think of magic as simply a form of fringe entertainment that deals with unique experiences rarely encountered during day-to-day life. However, as we have seen in this chapter, magical beliefs play an important role in our everyday cognitive processes.

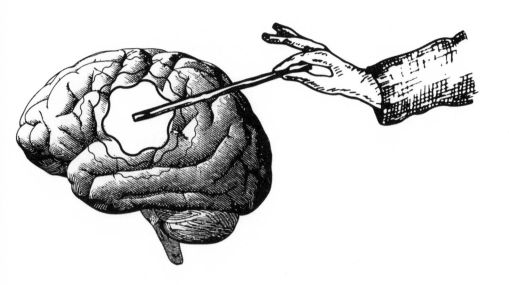

THE GAPS IN OUR CONSCIOUS EXPERIENCE

IT WAS A WARM SUMMER DAY in the early 1990s when first I saw Jim Cellini perform. Cellini was the best magician I had ever seen; he enchanted his audience with wonder, wit, and charisma. He was a short man with a distinct horseshoe moustache, who always performed wearing a black tailcoat and bowler hat. Although his performances were better than most Broadway shows, Cellini chose the street as his stage. He typically started his show by playing with a ball, which—to his own surprise and amazement—suddenly vanished, only to reappear behind a child's ear. Children were quickly mesmerized by these simple and funny tricks, and within a few moments, a substantial crowd would gather around him. He then performed more complex tricks, in which ropes were cut into pieces and magically restored, a lit cigarette vanished and reappeared, and smoke inexplicably started pouring from his mouth and ears. Cellini was the king of busking; he traveled the world performing for people on the streets.

Busking is a tough life: your wages depend on the coins that people toss into your hat, and the amount you get depends greatly on the strength of your final trick. In Cellini's case, this was always the Cups and Balls routine. This involved three brass cups and three small balls—and, of course, his magic wand. The audience was invited to examine the balls and cups to ensure they were fully legitimate. Cellini then made the balls vanish into thin air, only for them to reappear underneath the cups. For the next five minutes or more, the balls would jump from cup to cup, penetrating their solid bottoms and magically reassembling under a cup that had been freely chosen by one of the audience members. Cellini's trick was magical, but it what made it particularly memorable was the sudden emergence of several lemons and other large fruits from under the cup.

At the age of fifteen, I was fortunate to become one of Cellini's pupils, and much of what I know about the Cups and Balls (and magic in general) comes from him. Over the years, he taught me this beautiful routine, which I still frequently perform today. I consider the Cups and Balls to be one of the most fascinating pieces of magic in existence.

The Cups and Balls routine is sometimes known as *acetabula et calculi*; some historians claim that it dates back to the days of early Egypt (2500 BCE).[1] There are Roman descriptions of jugglers performing the Cups and Balls more than two thousand years ago, and it was also a popular trick in the Middle Ages. Hieronymus Bosch's late fifteenth-century painting *The Conjurer* provides some important insights into earlier versions of this routine (figure 4.1). The painting depicts a juggler performing the Cups and Balls for a small but fully engaged audience, misdirecting their attention to create an apparent miracle. And the deception does not stop there: a small boy exploits the misdirection to pick the pocket of one of the unsuspecting observers.

How does this trick work? It is not my intention here to give away the secrets of performance magic. But I can tell you that I simply put the balls into the cup and then let the rest of it happen right in front of your eyes—no trap doors, mirrors, matter transporters, or other secret devices. I simply misdirect your attention and then exploit your limited awareness to secretly load the lemons under the cup.

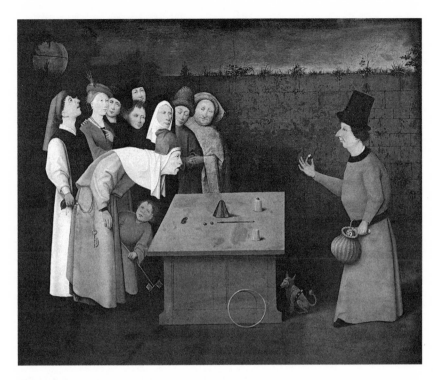

Figure 4.1

Cups and Balls: Hieronymus Bosch, *The Conjurer* (ca. 1475–1480)

From a magician's perspective, the Cups and Balls routine embodies almost every possible magic effect. From a scientist's perspective, it is a powerful illustration of our perceptual limitations. Indeed, it illustrates not one but two remarkable things about the mind. First, Bosch created his painting over five hundred years ago, at a time when our understanding of the human mind was rudimentary, to say the least. Yet his jugglers already knew that attentional misdirection can impair people's visual awareness. Second, and perhaps most importantly, the Cups and Balls routine illustrates just how wrong our intuitions about perception can be. We do not always see everything that happens in front of our eyes, even when we think we do.

So why can't we accept that magicians simply load lemons under the cups? Why are we willing to instead entertain the belief that the lemons appear through magic? The answer to these questions lies in the fact that one of the most powerful tricks our mind plays on us is making us believe that we are fully aware of our surroundings. It is a very compelling illusion—one that I will now try to shatter.

SHATTERING THE ILLUSION

Please take a break from reading and look around. What is your visual experience like? Mine is one of a world without gaps, a world filled with rich sensory detail, and this is likely the case for you too. The notion that your experience relies upon a detailed and complete representation of the world that is stored somewhere in your brain is very compelling; this has been an underlying assumption of cognitive science for decades.[2] Intuitively, your conscious mind feels like a private theater, with your conscious self located somewhere inside your brain, looking out at the world through your eyes. This is known as the *Cartesian theater*, a term coined by the philosopher and cognitive scientist Daniel Dennett.[3] The original idea goes back to Descartes's theory of dualism, which tries to explain how the mind interacts with the physical body. Descartes claims that conscious experience requires an immaterial soul and that the soul uses the pineal gland to interact with the mortal body.

There is no scientific evidence for the existence of the soul or for the pineal gland being the seat of consciousness. Nevertheless, Descartes's theory has been immensely influential. According to Dennett, many current theories of consciousness still rely upon the Cartesian theater. Most of these theories conceptualize a homunculus (a little man) who physically performs the task of observing all of the sensory data that is then projected onto an imaginary screen. This homunculus is the person responsible for making conscious decisions and sending commands to other parts of the brain and body.

Another common metaphor for consciousness is the flowing stream. The grandfather of modern psychology, William James, coined the phrase "stream of consciousness," and he suggested that the conscious self often feels like a continuously flow of thoughts and perceptual experiences. Susan Blackmore has suggested that few ever question this representation of consciousness because it seems so natural and intuitive.[4]

Indeed, as Dennett points out, the entire notion of a unified conscious experience might be an illusion in itself. Although we experience the world as a complete and detailed picture in our heads, our mental representations are much less complete than we think—they contain several large gaps, with many of the details in these gaps simply being filled in. One of the key principles in magic involves exploiting these gaps, and because it's hard to imagine that they exist, we rarely suspect that we have been tricked.

HOW MUCH OF THE WORLD CAN WE ENCODE?

Gaps in our perception of the world are present from the moment we begin picking up sensory information. For each of our senses, this involves registering a physical stimulus and transforming this information into neural signals, which are then processed by the brain. In the case of sight, physical objects reflect light, and our eyes register and encode its intensity and wavelength at every point in the visual field.

But our eyes only register electromagnetic radiation in the range of 340–740 nanometers; any light outside this range won't stimulate the receptors and thus won't be seen. The gap between perception and reality can be significant, as our inability to register something does not mean it's not there. For example, bees are sensitive to electromagnetic radiation in the ultraviolet range (where wavelengths are shorter) and thus can see patterns in the world that are invisible to us. This all might seem rather abstract, but several powerful magical techniques exploit this limitation. One of these is black light theater. Here, magicians perform in a dark room, selectively making particular objects visible and invisible.

Imagine that you had some magical paint that could turn invisible light visible. This is the actual situation for fluorescent colors, which absorb ultraviolet light and reemit it at a longer wavelength that can be seen by the human eye. In a dark room illuminated only by ultraviolet light, objects with fluorescent colors will glow. Covering such objects with a simple black fabric will render them completely invisible; uncovering them will make them visible again. Magicians often use this as the basis for stunning illusions.

Black light theater illustrates how easily our perceptual limitations can be exploited to create magical effects. But there are even more powerful and surprising ways in which magicians can manipulate our experience of the world.

HOW MUCH DO WE REALLY SEE?

The view that our mind holds a complete, detailed representation of the world—even of just the part visible to us—is difficult to challenge. So here is a quick thought experiment: have you ever experienced an object suddenly appearing in front of you? I'm not talking about how magicians make objects appear but simply how you might suddenly notice a cup of coffee that has been on your desk for the last week. You have spent days ignoring it, but once you notice it, you become fully aware of it. The cup was there the whole time, but were you conscious of it before you noticed it? Back in 1890, William James suggested that before an object is noticed, it simply does not exist in your conscious experience, and given that you don't seem to have gaps in your visual experience, your brain might simply be filling them in.[5] Could this be true? I would like to discuss this using one of the most amazing gaps that we all have, one that occurs when our receptors register incoming light: the blind spot.

Our eyes are designed in a very curious way. Incoming light falls on the retina, a network of millions of photoreceptors at the back of the eye. At some point in history, our ancestors developed simple eyes in which the neurons carrying information from these receptors travelled toward the center of the eye before heading to the brain.[6] Natural selection works with what is available, so these primitive eyes gradually evolved into complex eyes with muscles, allowing them to move; lenses, allowing them to form sharper images; and thousands of tightly packed photoreceptors, allowing them to encode visual information in high definition. By this time, the neurons carrying sensory information to the brain started to get in the way of the light, but it was too late to change the design. Evolution cannot decide that it has taken a wrong path and go back to the drawing board. Consequently, we ended up with imperfect eyes, and the nerves that transmit sensory information from the receptors therefore obscure some of the light that falls onto the retina. To make matters worse, these nerves all converge in a big bundle called the optic nerve, which exits the retina via a sizeable hole (an approximate six-degree visual angle in diameter) in which there are no receptors at all. This means that we are totally blind in part of our eyes. But curiously, we are rarely aware of this blind spot (see figure 4.2).

Part of the reason that we don't notice this is that we have two eyes, which can compensate for each other. Even if you close one eye, however,

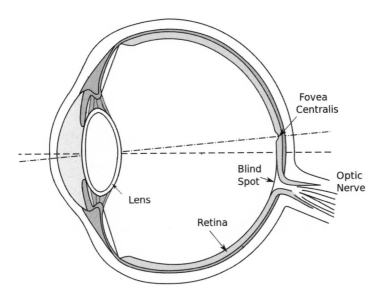

Figure 4.2

Diagram of the human eye, highlighting the fovea and the blind spot

Figure 4.3

Blind spot illusion: Close your left eye and keep your right eye fixed on the cross. Hold the book at arm's length and move it backward and forward until the top hat disappears.

you still won't notice this blind spot. But trust me, it's there! Let me prove this using a quick demonstration: Close your left eye and stretch the book out at arm's length. With your right eye, look at the small black cross in figure 4.3. As you slowly move the book backward and forward, you will notice that the top hat suddenly disappears. (This should happen about fifteen degrees out from where the cross is.) The top hat disappears because once it falls into the blind spot, you simply cannot encode the information from it. As far as your eye is concerned, it no longer exists.

The blind spot is a powerful illustration of how you fail to notice gaps in your visual experience, but it does not explain why you don't notice the magician loading the lemons into the cup. However, as we will see next, there are even more powerful examples of limits on your visual experience—limits which can be exploited by misdirection.

THE SCIENTIFIC STUDY OF ATTENTIONAL MISDIRECTION

My journey into the science of magic was almost accidental. It was largely the result of drinking vast amounts of coffee with Benjamin Tatler, who was studying how our eyes capture information from the world around us. Ben was working with Michael Land at Sussex University, where they built an eye tracker that allowed them to measure people's eye movements while interacting with the real world.[7] In most previous research on vision, subjects had simply interacted with flat virtual worlds rather than doing real-world activities, such as driving or making a cup of tea. Land is a true genius and maverick scientist who rarely shies away from a difficult scientific challenge—he even measured eye movements in mantis shrimp.[8] He also measured the eye movements (human ones) that occur during that most British of activities: making a cup of tea.[9] This work has been instrumental in helping us understand visual perception in the real world.

I was doing my PhD in a completely different area at the time, but their fascinating research made me wonder whether we could use an eye tracker to investigate misdirection. As a magician, I knew how to manipulate people's attention, and I came to realize that magicians had been exploiting many of the cognitive mechanisms that visual scientists were starting to study. After several cups of coffee, Ben and I embarked on the first scientific study on misdirection: we measured people's eye movements while they were being misdirected by a magician.

As part of this, we developed a special trick in which misdirection was used to prevent people from noticing a very obvious method. This trick was never intended to baffle or amaze; we simply wanted to understand the psychological processes exploited by attentional misdirection. I never thought that this misdirection trick would initiate so much new research. In the first version, I used misdirection to make a cigarette and a lighter disappear (see figure 4.4).[10] The method was simple: I sat at a table opposite the observer and picked up a cigarette, which I put in my mouth the

Figure 4.4

Misdirection trick used in Kuhn and Tatler, "Magic and Fixation."

wrong way round. I then used my other hand to pick up a lighter and pretended to light the cigarette, but I suddenly noticed that the cigarette was the wrong way round. My one hand then went up to my mouth and took the cigarette between my fingers, while my other hand (which was holding the lighter) moved down toward the tabletop and secretly released the lighter so that it fell into my lap. I then moved the empty hand up to my line of sight, snapped my fingers, and—hey, presto—the lighter disappeared. The lighter vanish acted as misdirection, because at exactly the same moment, my right hand let go of the cigarette, which visibly fell into my lap. Crucially, I made no effort to hide this drop; it was intended to be fully visible. After a few seconds, I raised my right hand and used a magical gesture to show that the cigarette had now disappeared. It is important to note that this is not a real magic trick, as magicians generally use much subtler ways of dropping an object. It was simply intended to study how misdirection can influence our visual experience.

We conducted this first study in the student bar because we wanted to test people's behavior in a natural environment. Ben set up the eye tracker, which allowed us to measure precisely where each person looked while they watched me perform the misdirection trick. To our amazement, only three of the twenty participants noticed the cigarette being dropped into my lap, even though this event seemed obvious. When we asked them to guess how the trick was done, most participants claimed that the cigarette must have gone up my sleeve. Although the cigarette had been dropped right in front of their eyes, they simply did not see it. When I repeated exactly the same trick, most participants immediately spotted how it was done and were flabbergasted that they had failed to notice it the first time round. Indeed, many refused to believe that I had performed this trick the same way both times.

Why did our participants fail to notice the obvious drop of the cigarette? At the time, we had several hypotheses. The first was that the surprise of seeing the lighter disappear, together with the sudden snap of my fingers, made them blink and so miss seeing the drop. Clearly, we are blind to incoming light each time that we blink. And as we saw in the previous chapter, people do indeed synchronize their blinking rates with times when the magician is more likely to carry out the secret method.[11] However, to our surprise, very few participants blinked at the time of the drop, ruling out this possibility.[12]

Another hypothesis was that participants moved their eyes at the time of the drop and so failed to see it. This hypothesis might need a bit more explanation: Our eyes continuously flitter about in an attempt to capture as much useful information as possible. Although we can move our eyes slowly to track moving objects, most of our eye movements—known as *saccades*—are very rapid; it only takes one hundred milliseconds or so for the eyes to move from one side of your visual field to the other. Taking a picture with a moving camera results in blurry images, and the same applies to the eyes. To avoid such images, our visual system simply suppresses input from the eyes during each saccade. Although you might notice yourself blink, it is impossible to notice these interruptions under normal circumstances. But try looking into a mirror while moving your eyes back and forth. You will not see them move; their movements occur during your saccades, exactly those moments during which you are blind. (You can see the movements of other people's eyes, of course, because your saccades usually do not happen at the same time as theirs.)

Although we are unaware of saccadic suppression, its effect is not small. David Melcher and Carol Colby calculate that the roughly one hundred milliseconds that we spend "offline" during each of the approximately 150,000 eye movements made each day add up to about four hours of blindness per day—one-quarter of our conscious life.[13] This figure is staggering. Our brain works hard to continually create the illusion of a seamless visual perception, one without this blindness.

Exploiting saccadic suppression clearly seems like a way to prevent people from perceiving a brief event. But in the case of the cigarette drop, observers rarely move their eyes during the drop, and there was no relationship between their saccades and whether they detected the event.[14] Moreover, in a different study, Anthony Barnhart and Stephen Goldinger showed that misdirection could prevent observers from noticing a coin that was simply pulled across the table.[15] Most participants failed to see this event, even though it occurred over a much longer duration than interruptions such as eye blinks or saccades. We therefore needed to start looking at other possibilities, ones that could create even larger gaps in our visual awareness.

WHAT ARE YOU LOOKING AT?

We generally assume that what we see is determined entirely by where we look. Indeed, everyday language considers the words "looking" and "seeing" to be pretty much the same. In this view, the reason why people fail to notice the method behind a trick is that they simply aren't looking at it. Indeed, many theories of misdirection claim that attentional misdirection is the manipulation of where people look.[16] This assumption makes a lot of sense, and it does correspond to what happens when our eyes begin to encode the information they receive from incoming light.

Intuitively, our eyes seem to work like cameras. Both devices have a lens that focuses light, followed by a sensor array that encodes the resulting pattern of energy. In digital cameras, this array consists of millions of tiny photodiodes evenly distributed across space; the total number of diodes determines the image resolution. In the eye, this array is the retina, a network of millions and millions of photoreceptors. Unlike the camera, however, the eye contains an uneven distribution of receptors; whereas the sensors in a camera are spread evenly across space, most of our color-sensitive receptors are packed into a very small part of our retina, known as the *fovea* (see figure 4.2). And although the fovea only covers about one percent of the retina, it projects to nearly fifty percent of the visual cortex. The fovea encodes information from a surprisingly small part of the visual field that amounts to roughly twice the width of your thumbnail at arm's length. Unless our eyes are looking at this very limited region, they simply will not encode information at high resolution or color. Most of the remaining visual field—the periphery—is therefore represented as blurry and black and white.

We are not entirely blind in the periphery; indeed, peripheral vision is highly sensitive to small changes in luminance and the onset of motion. But our peripheral vision cannot analyze fine detail, such as the words printed on this page. While reading, your eyes fixate on each word, so that it falls within the fovea. To demonstrate how this works, we can use eye trackers that allow us to create gaze-contingent displays in which the letters of the word being looking at are replaced with Xs. As an outside observer, it looks as though very little has changed because only one word at a time has been altered. However, because our peripheral vision lacks the resolution required to identify the individual letters, it makes reading almost impossible.[17] But if you reverse the manipulation and change all of the words apart from the one being looked at, people are perfectly capable of reading and barely notice that something strange is going on.

Even though I am experiencing my visual environment as if it included full sensory detail, the actual sensory information that my eyes are providing at each point in time is far less complete, because anything that does not fall within the fovea is blurred and coarse. As is apparent from figure 4.5, there is another huge gap in our sensory information, though this gap is more graded in nature. This extremely uneven distribution of visual processing resources raises some important questions. Why do we fail to notice our tremendously poor peripheral vision, and why is it so meager in the first place?

The human brain uses more energy than most other organs; although it only accounts for 2 percent of our total body mass, the brain uses nearly 20 percent of the overall energy consumed.[18] Although we still do not know exactly how this energy is used, approximately two-thirds of it is required to help nerve cells fire and is therefore directly involved in sending and processing information.[19] Processing information requires neural resources; the more information you want to process, the more neurons you require. These neurons not only require energy, they also need to be housed somewhere. Many of our technological advances have enabled us to create smaller and more powerful information processors. For our brain, there are severe physiological limits that determine how closely neurons can be packed together, and these limits have now been

Figure 4.5

The top image represents our visual experience. The bottom image represents the visual information our eyes capture during the course of four fixations.

reached. The only way to further enhance our processing power would be to grow bigger heads, which our bodies might simply fail to support.

The visual cortex takes up nearly one-third of the cortex, and visual processing comes at a great cost. If we processed all the information that falls upon the retina, we would need giant heads. Instead of humongous brains, however, evolution has driven us down a road of economy and efficiency. Rather than processing all of the high-resolution information that falls on our eyes, our visual system has our fovea pick up high-resolution information only from areas in the environment considered important; everything else is processed at a much lower resolution. Although we rarely think about where we look, we automatically fix our fovea onto aspects of the world that are of importance. These systematic eye movements are crucial to reading, and they are also fundamental to how we process information about our surroundings.

In his seminal work, Russian psychologist Alfred Yarbus used an eye tracker to measure where we look while viewing various kinds of pictures.[20] He discovered that we primarily look at those parts of our surroundings that are important. For example, if a picture includes people, we typically prioritize faces and, in particular, the eyes (see figure 4.6).

Given that we rarely think about where to look, how do we know what to attend to? Sometimes, we consciously decide what to focus on. For example, when making a cup of tea, you typically look at the objects that you are handling (e.g., tea cup, kettle).[21] But certain features or events in our surroundings automatically capture our attention; this is known as bottom-up control. Such control is best understood from an evolutionary perspective. We need to process information quickly and efficiently in order to respond to dangers swiftly. A delay in noticing a predator or ducking a fast-moving object could mean the difference between life and death. Thus, any mechanism capable of alerting us to potential danger is extremely advantageous, especially if this can be achieved efficiently.

As I sit at my desk, I can just about see the door of my room in the corner of my eye. At the moment, there is nothing to be gained from processing information about the door; it simply drains my cognitive resources and prevents me from coming up with the next sentence. But important things could happen in that doorway; a system that could alert me to people entering the room would be extremely useful. Rather than wasting cognitive resources on continuously processing door information, I

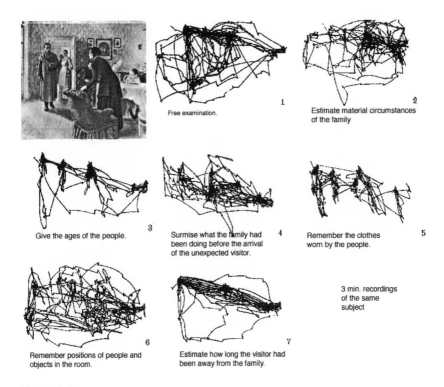

Figure 4.6

Early eye-tracking data that shows how eye movements are influenced by what you are doing (from Alfred Yarbus's *Eye Movements and Vision*)

could dedicate all of my resources to writing and only process information from the door when there is a high chance that something important will happen there.

This is exactly what our attentional system does: it uses clever tricks to alert us to anything in our environment that is potentially important. For example, our brain assumes that anything that suddenly moves is worth investigating. As soon as the door starts to move, there is a high chance that someone will enter the room. Similarly, as you are driving down a busy road, there is no need to process information about the parked cars until there is a sudden movement. This could be a car pulling out in front of you or a cat crossing the road. Detecting a simple perceptual change (e.g., change in luminance or color) requires far less time and computational resources than recognizing a cat. Because there is a high correlation between such perceptual changes and things that are important, it is far more efficient to process only the information in the area of a perceptual change than to process everything. Our visual system, therefore, continuously monitors for perceptual changes; once they have been registered, we attend to them and process the information more fully. Moreover, rather than building detailed representations of all aspects of our surroundings, we only represent objects and events when they are actually needed.[22]

Relying on an automatic process does have its downsides. Many of you will remember Clippy, the pop-up assistant who infuriated millions of Microsoft Word users with his hints and word-processing tips. As you were trying to formulate your next sentence, Clippy would pop up on your screen, wink at you, and in some cases even tap your monitor screen. The big problem with this was that Clippy always captured your attention; you simply could not ignore him. Clippy quickly became one of the least popular Microsoft Office features and was eventually killed off in 2011, in what was probably the most celebrated corporate assassination in history. But although Clippy has now gone to that great trash can in the sky, his spirit, so to speak, is still present: advertisers continue to create brightly colored, animated pop-up ads that automatically grab our attention and make us look at them while we are surfing the internet. It is often hard to ignore these ads because they exploit our automatic attentional control mechanisms.

Clippy had another characteristic that made him effective at grabbing our attention: his cute eyes. We are social creatures, and much of our success relies on our ability to cooperate with other humans. This in turn relies on making judgments about another person's desires and intentions, a process greatly facilitated by reading social signals. A person's facial expressions can provide valuable feedback about their emotional state, and their eyes can offer information about their desires and intentions. For example, we often use our eyes to look at objects that we are interested in. In most social interactions, it is very likely that things that are of interest to others will also be of interest to us. Consequently, attending to objects that are being looked at by others is a valuable attentional strategy.[23]

Our attentional system is finely tuned to processing social information, and these social attentional strategies develop very early in life. For example, newborn babies typically orient their gaze toward faces and, by about one year old, will start looking at objects that are looked at by others.[24] Adults too show a strong tendency to look at faces and will automatically follow the gaze of others.[25] Indeed, adults will follow another person's gaze even when they are explicitly warned not to do so.[26]

Social cues play an important role in magicians' misdirection, and we have conducted much research investigating how magicians use their eyes to control where you look and what you see. For example, in some of our early studies, we found that people generally follow the magician's gaze and typically look at the objects that he is looking at.[27] In the misdirection trick, the magician's gaze was used to misdirect attention away from where the object was being dropped, but we also conducted studies in which the magician directed his gaze toward the dropping object. We created two versions of the misdirection trick in which the magician's gaze was used either to misdirect attention away from the dropping object or to direct attention toward it.[28] People were four times more likely to notice the drop when the magician looked at it, and they were also more likely to look directly at it. These results illustrate that eye gaze is an extremely effective form of misdirection. Indeed, Andreas Hergovich and Bernhard Oberfichtner have come to the same conclusion by measuring participants' eye movements while they watched the Cups and Balls.[29]

Eye gaze is important in orienting attention toward different spatial locations, but the eyes themselves are also quite effective at grabbing attention in their own right. For example, using an eye-tracking study, we showed that people struggle not to look at a magician's face when asked a question.[30] Even though participants were explicitly told to keep their eyes on the magician's hands, more than half of the participants could not prevent themselves from looking at the magician's face once he looked at them. We also measured people's eye movements while they watched an entire routine of me performing the Cups and Balls routine. We found that their gaze was drawn to the magician's face once he looked at them.[31]

There are many good reasons to believe that people fail to see things simply because they do not look at them. Indeed, many theoretical models of vision presume that our conscious experience directly results from what our eyes look at and encode.[32] Thus magicians often assume that misdirection is about misdirecting where you look.[33] The scientific research on misdirection, however, does not support this view, and much of this work is changing our understanding of perception (see figure 4.7).

For example, Barnhart and Goldinger examined a misdirection trick in which a coin was pulled across a table. There was no difference between those who saw it and those who missed it in terms of where they were looking at the time the coin started to move.[34] Similarly, Tim Smith and colleagues used sleight of hand to misdirect participants from noticing a coin change; more than half of the participants failed to notice when a fifty-pence coin turned into a ten-pence coin, even though they were directly looking at the coin.[35] We found similar results using our color-changing card trick: participants were just as likely to miss noticing how the backs of the cards visibly changed from blue to red when they were looking at the cards compared to when they were looking at the magician's face.[36]

All of these results consistently show that where you look has an important impact on what information your brain receives. But looking

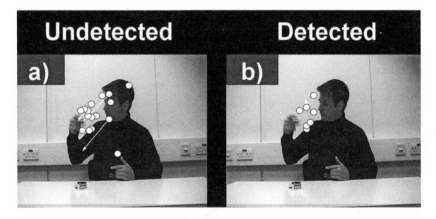

Figure 4.7

The figure depicts people's eye positions at the time when the cigarette was dropped, and we have divided participants between those who saw it and those who missed it. If participants missed seeing the drop because they were simply not looking in the right place, we would expect to find differences in eye positions between the two groups. However, this is clearly not the case, because the participants who detected the drop (b) were looking at similar locations to those who missed it (a). We even have one participant who is looking directly at the cigarette, yet he still missed seeing it. Data based on Kuhn et al., "Misdirection in Magic."

is not enough to see something. For that, you must also attend to that object, and this form of attention can be independent of what you are looking at.

WHAT ARE YOU ATTENDING TO?

Once your eyes have captured visual information from the world, countless other attentional processes along the visual processing stream must prioritize that information and filter out the less relevant aspects. Although eye movements play a central role in this, other attentional processes are also involved—processes that take place deep inside our brain.

Most early attention research focused on auditory attention. Imagine you are at a cocktail party or some other social gathering, where lots of people are talking simultaneously. To make matters worse, there is also a lot of background noise, but you can still listen to what your friend is saying while ignoring all the background chatter. You can also switch your attention to another conversation, although this prevents you from following what your friend is talking about. In the 1950s, in some of the earliest work on attention, Colin Cherry confirmed that people can indeed attend to only one stream of conversation at a time and can switch their attention to another one, but doing so prevents them from processing the meaning of the unattended stream.[37]

In 1999, Daniel Simons and Chris Chabris published a paper that transformed people's appreciation of attention.[38] They replicated an earlier experiment by Ulric Neisser that demonstrated the cocktail party effect in the visual domain.[39] Simons and Chabris presented people with a short video clip in which two teams of basketball players passed a ball from one player to another, and they asked participants to count the number of passes. Halfway through the experiment, a person dressed in a gorilla suit walked through the middle of the game. However, fewer than half the participants noticed this. The gorilla was obvious, and when participants watched the video again, they all were surprised to have missed it in the first place. This phenomenon—known as *inattentional blindness*—is so surprising that the gorilla video went viral and has since become one of the best-known psychological studies.

Inattentional blindness highlights the crucial role that attention plays in determining what we see.[40] Various studies have now shown that if people's attention is focused elsewhere, they can fail to see things even when directly looking at them.[41] Crucially, eye-tracking studies have shown that people looking at the gorilla are just as likely to miss it as those looking elsewhere.[42]

Another equally surprising phenomenon that highlights the gaps in our mental representation is known as *change blindness*. It was first encountered during an investigation into how we read.[43] In this study, participants failed to detect changes in word fonts (e.g., lower- to uppercase) if these changes were made during a saccade. John Grimes then discovered that this also happened when such changes were made to an image, even if the changes were large and seemed obvious once noticed.[44] But because the eye trackers and real-time computers needed for this were expensive at the time, these findings were hard to verify.

A few years later, Ron Rensink found that such failures occurred not only for changes made during an eye movement but whenever the change was not attended.[45] Rather than changing the image during a saccade, he simply alternated between an original and a changed image, with a brief blank inserted between the two images. This blank masked the visual transition that would normally have appeared at the location where the change occurred. This flicker paradigm cycled the changing images, with participants simply asked to say when they noticed the change (see figure 4.8). Intuitively, this task seems quite simple, yet it takes people ages to spot the change. But once the change is noticed, it is hard to imagine how you missed it in the first place. This is because you can now direct your attention to the object almost immediately and fully process it.

Change blindness can occur in many ways, such a making the change during an eyeblink or a brief occlusion of the changing object.[46] It is remarkable how blind we are to changes in our environment. For example, Daniel Levin and Daniel Simons carried out a fun experiment in which an experimenter approached an unsuspecting person on campus and asked her for directions.[47] After they chatted for a while, two men carrying a door walked between them, and the original experimenter was replaced by one of the men carrying the door. Even though the woman had clearly looked at the person, she failed to notice that she was now talking to an entirely different person. There are countless experiments that illustrate how inattentive people are to major changes in our environment.

Figure 4.8

Change blindness and the flicker paradigm

Change blindness explains why so many continuity errors go unnoticed.[48] Even though we often experience a film as a continuous scene, it has typically been cut together from different takes. Although directors exercise extreme care to ensure that each take is as similar as possible, errors often occur. For example, a half-empty glass can replenish itself from one cut to the next, or a shirt can magically unbutton itself. Once spotted, these errors cause much amusement, but what is more surprising is that most of them go completely unnoticed. Again, once you note an error, it's hard to believe that you could have missed it in the first place.

Change blindness and inattentional blindness demonstrate the massive, surprising gaps that can exist in our conscious awareness. Indeed, Levin and his colleagues found that most people fail to recognize how bad their perception can be.[49] They asked psychology students if they would have spotted some of the changes described in the experiments above. Most claimed that they would have readily noticed them. However, even though 98 percent of the people surveyed claimed that they would notice when a real person was switched with another, in reality only 46 percent did so. These findings are important in the context of magic because they illustrate how wrong we are about how little we consciously perceive, which is key to misdirection.

The reason why you don't see the magician loading the lemons under the cup or see how the cigarette is being dropped into my lap is because your attention is elsewhere, rendering you effectively blind. Instead of giving you a task such as counting basketball passes, the magician simply guides your attention away from the method used.[50] There has been much debate about the differences and similarities between misdirection and inattentional blindness. Nevertheless, it is clear that manipulating where, when, and what you attend to can result in astonishing failures in perception.[51]

Inattentional blindness also helps explain some of the key principles of magic. One of the golden rules states that you should never perform the same trick using the same method twice, and people are indeed far more likely to notice the cigarette or lighter drop when the same trick is repeated.[52] The reason for this is that seeing a magic trick for the first time requires more cognitive resources, and so you have less attentional capacity available to see the method. Similarly, telling your audience what you are about to do is a bad idea; our studies show that once participants know that I am about to make a cigarette and a lighter disappear, they are significantly more likely to notice the drop.[53] The reason for this is that once you know what the magician is about to do, you can free up attentional resources and thus be more likely to spot the secret method.

WHY DOES IT MATTER?

By the early 1960s, the dominant approach to understanding the mind was cognitive psychology. This approach is based on information processing, and many of its early theories used computer analogies to describe how the mind works. For example, an input device such as a keyboard or camera was used to symbolize our senses, while printed outputs and robot actions corresponded to conscious thoughts and motor responses, respectively. Like us, computers use different memory devices with different functions and properties. For example, RAM has a low storage capacity but rapid information access, while hard drives are more sluggish but can store much more data.

This is how I learned to think about the mind as an undergraduate, and such analogies can still be encountered today. However, many of these ideas are as outdated as the desktop PC I had at the time. Our brains are much more like a smartphone than a 1990s PC. Smartphones access information when they need it; there is no need to store the information on the actual device, which makes them very efficient. Imagine that you knew nothing about the internet and I gave you my iPhone to check the train timetables or read yesterday's paper. Given that it displays all of this information, you would automatically assume that my phone must have a massive memory, but this is an illusion.[54] The iPhone simply downloads the information from the web whenever it is needed, but you have no way of knowing this.

Our conscious experience works in a similar way: rather than processing and storing all the information that falls on our eyes, we simply process and remember things as we need them. If we can access this information whenever we need to, there is simply no need to store it all in our mind. As with the smartphone, you don't notice that the information is not stored in your brain, and you don't notice the gaps in your conscious awareness because your visual system provides you with the necessary information as soon as you need it.

Our failure to appreciate just how little we consciously see lies at the heart of magic. This failure also has wide-ranging practical implications. Back in 1980, scientists from the National Aeronautics and Space Administration (NASA) carried out research on heads-up display technologies, which project information from airplane instruments directly onto the windshield in front of the pilots.[55] These displays were intended to help pilots spot potential dangers outside, allowing them to read instruments without having to take their eyes off the flight path. Eight highly experienced pilots were trained to use these displays in a flight simulator and were then tested to see if this new system would help them notice potential hazards. But a rather surprising result was found: as the plane broke through the clouds and approached landing, several of the pilots completely failed to notice a large jet turning onto the runway directly in front of them. The pilots were surprised to learn that they had failed to see the huge jet right in front of their eyes.

Such dramatic failures in perception are shocking, but now that we know about the massive gaps in our conscious awareness, the reason for them should be clear: Even though the pilots' eyes were in the right place, their attention was not. Consequently, they simply did not see what was in front of them.

Similar failures can also occur in our own lives. We live in a fast-paced world where our attention is continuously distracted, and despite thinking that we are capable of multitasking, we are not (another example of how we fail to acknowledge our cognitive limitations). This can have problematic consequences. Ira Hyman and colleagues investigated how conversing on a mobile phone influenced participants' walking behavior and their ability to notice unexpected events.[56] Talking on a mobile phone significantly slowed down their walking speed, but more importantly, it also significantly reduced their chance (25 percent) of noticing a brightly colored clown riding a unicycle, compared to participants who were either listening to music (61 percent) or simply walking (51 percent).

Missing the clown points to a more perilous consequence for drivers. People typically find it difficult to believe that mobile-phone use impairs their awareness, which is one of the main reasons it is so hard to convince motorists to not use them while driving.[57] Numerous studies using driving simulators have shown that talking on the phone results in poor driving performance—poorer than when listening to music or audiobooks or when conversing with a passenger.[58] As it turns out, engaging in a mobile-phone conversation while driving is even more dangerous than driving while intoxicated.[59]

The United Kingdom and most US states have outlawed the use of handheld mobile phones while driving, with motorists encouraged to use hands-free sets instead. The rationale for this is the erroneous assumption that mobile phones impair driving because motorists look at their phones and so take their eyes off the road. However, we are beginning to realize that attentional distraction plays a far more important role in determining what we see than does where we look. Numerous studies have now demonstrated that people drive just as poorly with hands-free mobile phones as they do with handheld devices.[60] (Distressingly, participants in these studies often felt that talking while driving did not impair their performance.) Note that talking to fellow passengers is less distracting, because they can—and typically will—stop talking when you need to pay attention to the road. Fellow passengers can also provide the driver with useful additional information (such as noticing a car approaching), which mitigates some of the bad effects. But recent research has shown that even the mere notification of having received a phone call, without even answering the phone, has a negative impact on our driving performance.[61]

WHAT CAN BE DONE TO ALLEVIATE THESE PROBLEMS?

Our attentional system evolved so that we could process information as efficiently as possible in our ancestral environment. But technological advances have resulted in an environment that requires more and more of our attention. This new understanding of our limitations will hopefully cause people to think more carefully about technological developments, as well as the judgments we make of others. For example, would it be a good idea to develop augmented reality devices that let us check email and social media while driving? Augmented reality interfaces are being developed to help surgeons perform endoscopic surgeries, and despite the fact that these devices have been shown to increase the surgeons' precision, nearly half of the surgeons using this device failed to notice a critical complication—a screw that was near their target point.[62] Advances in future interactive display technologies have great potential, but they are only safe if they accommodate our perceptual limitations.

Awareness of our perceptual limitations also has important implications for the legal profession, where people struggle to distinguish between ordinary perceptual failures and situations in which individuals have been negligent.[63] For example, it is easy to see how a police officer who missed seeing a crime could be accused of negligence. Yet as we have learned here, this could simply reflect a general perceptual failure.[64] Indeed, a study by Chris Chabris and colleagues showed that most participants failed to notice a staged fight that occurred in plain sight within ten yards of them.[65]

Intuitively, we feel like we are aware of our surroundings, yet as we have seen here, huge gaps exist in our conscious experience. And it is the counterintuitive nature of these limitations that lies at the heart of magic. Now that we have learned about just how little we actually see, we will tackle the question of whether the things we consciously perceive are in fact as we believe them to be. We may trust our eyes, but as we have learned, there is much more to vision than meets the eye. In the next chapter, we will explore the validity of the things we consciously perceive and will discover that perception is more subjective than you might think.

SEEING IS BELIEVING

DAVID COPPERFIELD IS ONE of the most successful magicians of all time, and the Death Saw is one of his most brilliant illusions. Standing on a huge stage, Copperfield proclaims that you are about to see an escape like you have never seen before. Members of the audience are invited to inspect padlocks, which will be used to strap him onto a table. Seventeen feet above this table, there is a circular saw measuring six feet in diameter and capable of cutting through an eighteen-inch concrete block. The descent of the saw blade is controlled by a timer, which allows him only sixty seconds to escape, otherwise he will be cut in half. Copperfield is strapped down to the table, and a box is closed around him, yet his hands and feet are clearly visible. As the countdown starts, the saw gradually descends, getting closer and closer to his body. The tension rises. By now he has managed to open the box, and we can see his entire body, but he is obviously struggling with one of the locks. Time is not on his side, and although he has freed himself from the shackle around his neck, a technical fault leads to a short circuit. In a dramatic twist, the huge revolving saw blade suddenly drops, cutting Copperfield in half. Everyone gasps, and we are left with the shocking image of a huge saw blade having severed his torso.

After a few intense moments, he lifts his head. His charming smile tells us that all is fine, much to our relief. Two assistants come onto the stage and pull the tables apart, revealing that his body has clearly been cut in half. How is this possible? Copperfield's body is in full view, and the table is far too thin to hide a second person. We know that this must be an illusion, yet no matter how hard we look, it is simply impossible to work out how it is done.

There is a long history of magicians sawing people in half, going back to January 17, 1921, when the English magician Percy Thomas Selbit first publicly sawed a woman in half at the Finsbury Park Empire theater in London. This illusion has become one of the most iconic magic tricks, only rivaled by pulling a rabbit out of a hat. I'm still frequently asked whether I have ever sawed my wife in half (the answer is no).

Magicians use the term "illusion" to describe large-scale magic performances that are typically performed on a stage. For magicians, the defining feature of an illusion is simply its scale. However, more generally, illusions lie at the heart of magic, and they play a fundamental role in many smaller magic tricks. In the previous chapter, we discussed the huge gaps in our conscious awareness and the illusion of being aware of our surroundings. We will now look at how our perceptual experiences are far more subjective than we think. As I look at Copperfield's body sliced in two, I simply cannot work out how this is possible, no matter how hard I look. Many of these illusions work because they manipulate the way we perceive the world. In this chapter, we will look into why our minds are so easily tricked into seeing something impossible. Doing so will require us to look deep into the processes of sensation and perception, and we will uncover some of the truly amazing tricks that our brain uses to perceive the world around us.

WHAT IS AN ILLUSION?

Vision is our most trusted sense, and our eyes provide valuable information about our environment. But can you really trust your eyes? Intuitively, we think of vision as a process by which our eyes take lots of pictures of the world, which are then stored and processed deep inside our brain. Even though we trust our eyes, the things we perceive may not necessarily be as truthful as we think they are. As a young child, my mum told me about an interesting perceptual conundrum. Why do we see the world in its correct orientation? This is a question few of us think about, but it's a question that lies at the heart of perception. Our eyes capture upside-down images, yet we perceive the world in its correct orientation. At the time, my young teenage brain was just about able to understand that our eyes have a lens that projects visual images onto our retina, and my optics physics set (yes, I was a geek) taught me that lenses invert the images. However, I simply could not grasp how our experience of the world could be so different from the information our eyes capture. Inverting images is actually one of the easier computational transformations that our visual system completes, because our visual experience is far more removed from sensory information than most of us imagine.

Our eyes do indeed take pictures of the world, but we never see these retinal images. Although the retina provides information for seeing, we don't actually see the information that it captures. Instead, our brain uses the images as a starting point to make sense of this visual information, which eventually leads to a conscious experience. Once the pictorial information has been captured, different parts of our brain encode and process different visual properties. For example, cells in an area known as the primary visual cortex are sensitive to edges, and they help us identify lines and corners. This information is useful for identifying the size and shape of an object. Other brain areas process color, and cells in a part of the visual cortex known as V5 are simply responsible for detecting motion. The picture-like quality of our visual experience makes it very difficult to imagine that our vision entails anything other than pictures. For example, as I'm looking out of the window, I see a cat walking across the grass, but my brain does not receive simple pictures of this moving cat. Instead, the visual cortex receives lots of different bits of information about the visual properties, and my brain then guesses what might be out there: it's a cat.

We intuitively believe that there is a direct relationship between our perceptual experience and the objects in the world, yet this relationship is far less direct than we think. Perception is all about interpretation, which means that we see objects depending on how our brain interprets them. This idea is far from new. In the mid-nineteenth century, the German physician Hermann Ludwig Ferdinand von Helmholtz proposed that perception is a process of unconscious inference, in which we automatically and unconsciously come up with a best guess about the structure of the world that is consistent with both the retinal image and our past experience.[1] This means that our trusted sense of vision is nothing more than informed guess work.

There are many perceptual illusions that highlight the subjective nature of our visual perception. Take a look at the figure below, depicting a drawing of a cube.

Figure 5.1 depicts a Necker Cube and belongs to a group of visual illusions known as bistable images. The figure contains no visual cues about the cube's orientation, which means that it can be interpreted as having either the upper-right square or the lower-left square as its front side. As you stare at this figure, your perception of the cube will flip from one version to the other, even though the physical image remains the same. This illusion illustrates that your perception relies on your own interpretation of the world around you, and thus your perception does not directly correspond to physical reality. Rather than an exact science, perception is a

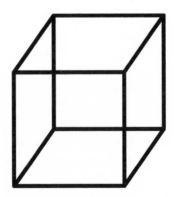

Figure 5.1
Necker Cube

cognitive process involving lots of guesswork that often goes well beyond the available evidence at hand.

Perception is a form of reasoning and hypothesis testing, which resembles how we reason about other things in our day-to-day lives. As you are reading this book, you are using past knowledge to interpret the meaning of each word, and we all know that words can take on different meanings. For example, if I tell you that a fisherman is walking along the bank, you will likely interpret the word "bank" as denoting a riverbank. However, if I tell you that robbers are about to steal money from the bank, the word "bank" suddenly takes on a very different meaning. The same is true for perception. Depending on how we interpret the visual signals, the same retinal image can lead to very different perceptual experiences.

It is difficult to imagine that our trusted sense of sight results from guesswork, but this becomes apparent once we start to look at how easily we can be tricked into seeing things that aren't actually there. While I am typing these words, I can clearly see that my hands are in front of the laptop screen and that my cup of coffee is within easy reach. Making judgments about the spatial relationship between objects seems like a trivial task and one that most of us take for granted. However, depth perception is far from simple. Our eyes capture flat, two-dimensional snapshots of the world, which contain no direct information about depth. Our perception of depth relies on clever tricks to extrapolate the third dimension.

Although depth cannot be perceived directly, there are lots of clues that give us indications about the potential spatial location of an object. For example, if your hand occludes the book you are holding, it is extremely likely that the book is farther away than your hand. There are many other pictorial cues that provide insights into the spatial relationship between objects. For example, if you are looking at a road that goes away from you, the road gets thinner the farther away it gets. This is because parallel lines converge at a distance, and these perspective cues are valuable for judging distance. We also exploit the fact that our two eyes see the same object from slightly different angles, thus resulting in slightly different retinal images. Our visual system uses a clever process of triangulation to compare the disparity between the two images and uses this information to calculate the relative distance of an object, a process known as stereopsis.

Seeing depth depends on piecing together lots of different clues, but there is no guarantee that the depth you are experiencing is indeed true, which means you can easily be tricked into experiencing something that is not necessarily there. There is a long tradition of fooling people with illusory depth. For example, in the Renaissance, painters discovered perspective and used artistic tricks to add the illusion of depth to previously flat, two-dimensional images.[2] The trick involved using pictorial depth cues that emulate the way in which three-dimensional objects are typically projected onto our retina. For example, drawing converging lines provides the illusion of seeing an object receding away from us. Similarly, if you draw one person smaller than the other, the smaller one is perceived as being farther away than the larger one. These pictorial cues are effective at adding depth to a two-dimensional image, but in most cases, you are perfectly capable of distinguishing between illusory depth and the real three-dimensional world.

In recent years, there have been huge technical advances in computer animation and projection, which means that the distinction between illusory depth and real depth has become more blurred. Most animated blockbuster films are now produced in 3-D, allowing you to experience illusory depth. The trick behind 3-D animation lies in presenting a slightly different image to each of your eyes. These images are constructed so that the offset between them emulates the binocular disparity that you would experience when an object is perceived at a distance. In a 3-D cinema, polarizing filters are used to project these two images onto the screen, with each image using a different polarity of light. The polarizing glasses worn by the audience contain a pair of polarizing filters that allow light of a particular polarity to enter the eye, while blocking light with a different polarity. These glasses can be used to present different images to your eyes. Our visual system can't distinguish between the retinal disparity produced by real three-dimensional objects and that produced by this clever projection system, which is why we experience the images in the 3-D cinema as having true depth. Virtual reality devices use similar tricks.

Most of us do not consider 3-D cinema to be magic, but back in Victorian London, two inventors came up with an optical illusion involving a clever projection system that is still commonly used in magic and theatrical performances today. The Royal Polytechnic Institution on upper Regent Street was an important feature of Victorian London and, similar to the present science museum, held classes on technical education.[3] In 1848, John Henry Pepper took charge of the interactive exhibits, and in 1862, he incorporated a new optical illusion invented by Henry Dircks, which today is known as Pepper's Ghost. The first public presentation of Pepper's Ghost depicted a medieval student, hard at work in his room. Suddenly, a ghostly skeleton robed in a shroud magically arose from the floor, groaning horribly. Driven to distraction, the student picked up a sword and made a wild thrust at the apparition, which immediately disappeared. But as soon as he returned to his book, the ghost was back at his side. When confronted, the sword passed harmlessly through his shrouds, after which he simply vanished.

This magical demonstration was a tremendous success, and the principle behind the illusion is as powerful now as it was back then. Pepper's Ghost relies on a very clever optical trick that exploits the way our brain interprets sensory information. A glass plate is placed diagonally across the stage, in such a way that it is both invisible to the audience and capable of acting as a mirror that reflects images on its surface. Glass can simultaneously function as a mirror and be transparent, which leads to a strange optical illusion. You can create this illusion yourself at night by turning off the lights and holding a burning candle in front of the window. Once you get the viewing angle right, you can see through the window while at the same time seeing the reflection of the flame, spatially located outside the room. Your brain interprets this sensory information as resulting from a flame that is levitating outside your window, which of course is impossible.

Professor Pepper's stage was set up so that the mirror reflected the contents of a hidden room, which was typically located in the wings of the stage. If this hidden room is dark, nothing is reflected in the mirror, but once it is illuminated, its contents are mirrored. From the audience's point of view, this reflection of the illuminated contents magically appears in the middle of the stage. Because the reflection is not perfect, the image is slightly translucent, which adds to the ghostly experience. Pepper's Ghost allows you to project images into empty space, and this principle is often used in amusement parks and theater productions. For example, the award-winning West End musical *Ghost* used this principle to conjure up its spirit, which left contemporary audiences gasping.

Pepper's original setup was quite cumbersome because it required a heavy, fragile glass plate, which is partly responsible for its diminished use today. However, advances in high-definition projection technologies have led to a revival of the illusion. The illusion no longer requires a hidden room and heavy sheets of glass. All you need is a thin layer of special cling film stretched across the stage. This acts as mirror capable of reflecting a high-definition image, which can be prerecorded or streamed live over the internet. Companies such as Musion specialize in using this illusion to create onstage holographic displays, generating images of people and objects that are barely distinguishable from reality. For example, this technique has been used to project an animation of deceased rap artist Tupac Shakur onto a stage, allowing him to perform with real, living musicians. The effects are truly breathtaking; I've seen these displays in real life, and it's the closest I have ever come to seeing a true hologram.

We have all had different past experiences, which influence our beliefs and knowledge about the world. Inasmuch as perception is based on our own past experience, all of us should perceive the world differently. Let us explore this point in more detail using one of my favorite perceptual illusions, the Hollow Mask Illusion.[4] As you look at the hollow mask depicted in figure 5.2, you experience it as being solid.

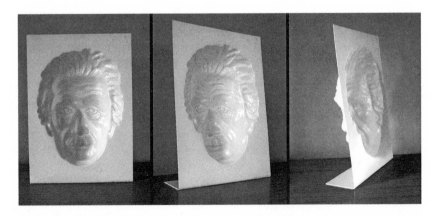

Figure 5.2

Hollow Mask Illusion

I have this Albert Einstein mask standing on my desk right in front of me, and even though I'm fully aware that the mask is hollow, my brain refuses to accept this, and I perceive the mask as a solid. Why is my brain tricked into perceiving a solid face? The Hollow Mask Illusion illustrates how we use past knowledge to generate perceptual experiences. We spend a lot of our time looking at faces, and the vast majority of faces we encounter are convex and thus have noses pointing outward. When presented with a face, you can usually safely assume that the nose is pointing outward, and this is exactly what your brain does. Although there are lots of other cues in the environment (e.g., stereo vision, shadows) that should tell you that this Einstein mask is hollow, your brain makes an educated guess, and you therefore see what you believe to be true.

If "seeing is believing," then people with different beliefs should perceive the world differently. There is much research illustrating that this is indeed the case. For example, one study has shown that infants with less experience in processing faces are less susceptible to the Hollow Mask Illusion and perceive the face for what it is: hollow.[5] Similarly, individuals with schizophrenia are thought to use less top-down knowledge to interpret what they see and thus perceive the world "more truthfully" than the majority of people. Indeed, individuals with schizophrenia accurately perceive the hollow mask as hollow.[6] Likewise, individuals with autism employ different visual processing strategies and have been shown to be less susceptible to a variety of visual illusions.[7] As we will see later on, however, individuals with autism are not necessarily more resilient to all types of magical illusions.[8]

We typically think of illusions as instances where we experience deviations between our perceptions and reality. However, as we are discovering in this chapter, all of our perceptual experiences are based on informed guesswork rather than objective reality, which means that all perception turns out to be an illusion. The idea that perception is simply an illusion is rather unsettling, and it also raises an important question about the usefulness of the term "illusion" itself. If everything is simply an illusion, what is the point of talking about illusions? According to Richard Gregory, the idea that all of perception is simply an illusion is as useful as claiming that everything is a dream.[9]

Although the term "illusion" may indeed be superfluous, the systematic perceptual distortions that we experience provide us with valuable insights into the cognitive mechanisms that underlie our perceptual experiences. This is because illusions do not result simply from "sloppy computations" but from statistically optimal computations that are functionally beneficial in the real world. As we will see next, many principles in magic exploit the computational tricks that the brain uses to make sense of the world. Magicians use them to manipulate your perceptual experience.

MAKING SENSE OF THE WORLD

Our visual experience is more subjective than we think, but how does this explain how a magician can cut a woman in half? To understand this illusion, we need to explore the way our brain tries to make sense of the visual world. We are very good at recognizing objects, but we rarely think about the computational complexities that lie behind these processes. Many of the tricks that our brain uses to perceive the world for what it is are also responsible for creating some of the most compelling illusions.

The first stage in object perception involves determining which parts of the image constitute an object and which are part of the background. Objects rarely present themselves in isolation, and we typically find them in cluttered environments. To make things worse, they are also often partially occluded by other things, meaning that we don't even get a full view. For example, my coffee cup is partially occluded by my laptop, and it is surrounded by lots of other visual clutter. Yet I have no problem in recognizing it. How does my brain know which parts of the visual scene belong to the cup and which belong to the laptop or other parts of the background?

It is much easier to segregate visual objects from their environment once we know what they are, but this presumes that I already know the object's identity. A system that only recognizes things that it already knows is fairly limited, and clearly our visual system can do more. We can look at things and use vision to discover what they are. Figure-ground segregation is far from easy, but luckily there are perceptual properties that provide clues as to which parts of an image belong together. For example, visual features that are similar (e.g., same pattern, same color) are likely to be part of the same object, and features that are in close proximity are also likely to belong together.

In the 1920s, a group of German psychologists became interested in understanding the way that objects are grouped and came up with a set of laws for determining the types of features that are grouped together. These gestalt grouping principles are important in influencing how objects are perceived. For example, the law of closure states that people perceive objects such as shapes or letters as being whole, even though they are not necessarily complete. The Kanizsa Triangle is a beautiful illustration of the power of this grouping principle. Although the "triangle" in figure 5.3 (left) is not complete, we perceive it as if it were intact. In fact, this grouping principle is so strong that we even experience illusory lines or contours between the missing segments. Moreover, as you look at the figure, you will also experience illusory luminance, in that the white in the triangle looks brighter than the background.

The law of continuation states that elements of objects are grouped together, and features that are aligned tend to be perceived as whole objects. Importantly, if there is an obstruction between the objects, people still perceive them as a single uninterrupted entity rather than two separate objects. This principle makes adaptive sense because our visual world is generally rather cluttered, and objects are often occluded by other things. However, just because an object has been occluded does not imply that part of that object has been removed. It's much more plausible that the object remains intact while out of view.

Gestalt laws provide us with valuable descriptions of how visual features are grouped. The beauty of these laws is that they provide low-level clues as to which features are grouped together to form an object. By

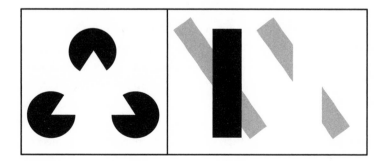

Figure 5.3

Kanizsa triangle (left); principle of good continuation (right)

"low-level," I refer to simple visual features such as lines, dots, or edges. With such clues to rely on, the observer does not require knowledge about what the object is in advance. As with most laws, they can be broken. So despite the fact that gestalt principles give us valuable estimations about the nature of an object, there is no guarantee that all objects behave according to gestalt laws.

For example, we perceive figure 5.3 (right) as a wider bar on top of a thinner one, but there is no guarantee that this is indeed the case. It is entirely possible that the thinner bar consists of two disconnected segments. As we have seen, perception isn't necessarily an exact science, and many different environmental configurations can produce the same percept. Gestalt principles are helpful for disambiguating lots of different scenarios, but this process is not foolproof and can lead to errors—perceptual errors that magicians frequently exploit.

Let me start by describing one of the first magic tricks I learned to perform when I was a kid. Here is what the audience sees: I reach into my pocket and reveal a jumbo-sized card depicting the four of spades on one side and the ace of spades on the other side. After mumbling some magic words, I turn the card, and to the audience's astonishment, the four of spades turns into the six of spades. I turn the card again, and now the ace of spades has turned into the three of spades. This is not the most impressive magic in the world, but it is a wonderful demonstration of how gestalt laws can be broken to produce magic tricks. So how is it done?

The method for this trick does not rely on sleight of hand. All you need is a gimmicked card that exploits our gestalt grouping principles. Although the audience sees four different cards (ace, three, four, six), there are only two. However, these two cards are not normal. One of them is a six of spades with one of the middle spades missing, and the other one is a three of hearts with the bottom heart missing. By strategically holding the card at different locations, the audience will see the card as being different. For example, if you hold the fake six by covering one of the central pips, people will see it as a four. However, if you hold it by covering the empty space, people will see it as six (see figure 5.4). The same principle applies to the ace and the three.

The reason this is so effective is that our visual system uses the principle of symmetry to interpret the visual world in terms of the most likely

Figure 5.4

Card trick based on visual symmetry resulting in amodal completion (left) and amodal absence (right)

object given the perceptual constraints. It assumes that objects are symmetrical, and the fact that one aspect of the object has been occluded does not imply that it is not there. In most cases, this is a useful strategy, as it prevents us from being confused by occlusions. However, when broken, this law leads to an illusion.

Ever since Uri Geller astonished the world with his mystical spoon-bending power, magicians have developed more and more sophisticated ways to make objects magically twist or change form. Indeed, changing the physical properties of objects is one of the fundamental effects in magic. Destroying objects is generally not that difficult and doesn't require much magic. The thing that makes it magical is that the conjurer magically restores the object to its original form. Many of these effects rely on exploiting our bias toward seeing interrupted objects as being continuous rather than disconnected entities.

The spoon-bending trick is a beautiful demonstration of how the law of good continuation can be exploited to create a magical effect. In this illusion, the magician picks up a spoon and holds it between his fingers. He then gently touches the bottom of the spoon and twists and bends the metal spoon as if it were rubber. In some cases, a gentle rub or blow is sufficient to turn metal into jelly. How is this possible? The secret lies in using two spoons rather than one. To be more precise, the magician uses one and a half spoons: one whole spoon and the bottom half of another spoon. The method behind this spoon-bending effect requires the magician to hold the two spoons so that they are perceived as being one. He does this by hiding the bottom bit of the whole spoon beneath his fingers, and then places the gimmicked half spoon on top of this. The key to this illusion is to cover the joint between the two spoons with your finger. As the spoon is now made up of two separate spoons, it is extremely easy to give the impression that you are bending it. Because of the law of continuity, observers generally perceive the interrupted spoon as one continuous spoon.[10]

Gestalt grouping principles also play a very important role in stage illusions. As some of you may have guessed by now, they also help to explain how magicians create the illusion of cutting a woman in half. There are countless variations of this effect. The most popular version, the Zig Zag Illusion, was invented by Robin Harbin back in 1960s. In this illusion, the magician's assistant climbs inside a box, which is then closed. Although most of her body is concealed, her head, hand, and foot are clearly visible throughout the illusion. The magician then pushes two blades through the box, apparently cutting the woman into three pieces. As depicted in figure 5.5, he then proceeds to slide the middle section to one side to further illustrate that the woman has been cut in pieces.

So how is it done? The illusion relies on two principles: First, the box is painted in a clever way that means it looks smaller than it actually is. This is largely achieved by painting the sides black and thus making the box appear much thinner. When the middle section is pushed to one side, the woman needs to contort herself so that she can fit into the side section of the box. However, the thing that makes this illusion particularly effective is the fact that people do not expect the woman to contort herself. This is achieved by giving observers visual clues implying that the woman has remained in the original position. For example, each of the three segments has a hole, through which the assistant puts one of her body parts. This means that throughout the trick, spectators can keep an eye on the head, hand, and foot. The second feature that really makes this illusion work is the fact that the front of the box contains a silhouette of a person. When the woman enters the box, the outline accurately describes her location in the box. However, as soon as the door is shut, the assistant contorts herself to squeeze into the side position, and thus the silhouette no longer

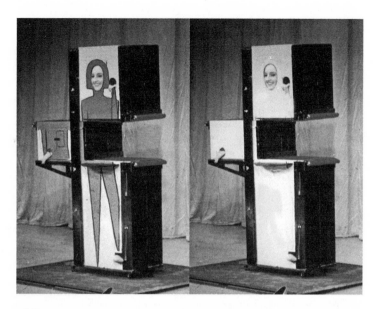

Figure 5.5

Zig Zag Illusion: A schematic outline of the human body enhances the effectiveness of this illusion (left). When the outline is removed (right), the illusion is less effective.

represents her actual body position. In our mind, the three body parts are clearly combined, and as soon as the continuity of the silhouette is disrupted, we believe that the impossible has happened. Anthony Barnhart has nicely demonstrated that removing the silhouette from the illusion reduces its effectiveness.[11]

In recent years, Vebjørn Ekroll, Bilge Sayim, and Johan Wagemans have started to investigate a wide range of magic tricks, and their findings are revealing impenetrable perceptual illusions that are far more powerful than most magicians would expect.[12] For example, the Chicago Multiplying Billiard Balls Routine is a classic in magic, and it beautifully illustrates just how compelling some of these illusions are. In this magic trick, the magician starts with one ball, held between two fingers, and we see that his hand is otherwise empty. All of a sudden, and apparently out of nowhere, a second ball appears between his adjacent fingers. Using the same principle, the magician can make other balls appear, which afterward magically vanish into thin air.

The secret behind this trick is that one of the balls is a hollow shell, from which an additional ball can be secretly produced. As the magician relocates one of the balls using his other hand, he secretly loads another ball inside the shell, which is produced at a later point. Once the secret has been revealed, the illusion suddenly seems rather disappointing. More importantly, it is hard to imagine that people would truly experience the half ball as a solid sphere. However, take a look at figure 5.6, which shows my hand holding a solid ball and a shell. I have performed this illusion many times, and although this trick cannot be performed with people watching from behind, I've always been surprised that people do not discover the secret. Ekroll's clever empirical research has shown that the illusion is far more compelling than anyone could have imagined.

The illusion persists even once you know how it is achieved, and your experience of seeing the back of the halved ball is more than just imagination. Ekroll and his colleagues have suggested that when "viewed from this perspective, the spectators do not merely entertain the intellectual belief that the semi-spherical shell is a solid ball, but rather automatically and immediately perceive it as such because that is what their visual systems tell them to be the case."[13] You can try the illusion yourself: take a Ping-Pong ball and cut it in half, then place it on a flat surface such as a table. When the ball is flat on the table surface, we perceive it as a solid

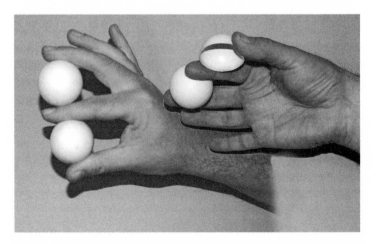

Figure 5.6
Chicago Multiplying Billiard Balls Routine

half sphere. However, if you start lifting the shell slowly upward from the table, you will get the impression that the semisphere is transforming into a full sphere. The amount of "growth" is proportional to the space between the sphere and the table. Once you start lowering the sphere back to the table, it starts to shrink again. Even more strikingly, balance this shell on your fingertip and look at it from above; it looks like the ball is solid. Most bizarrely though, Ekroll and his colleagues have shown that now your finger appears to have shrunk.[14] Ekroll explains that the mere thought of a halved Ping-Pong ball as being complete is unlikely to make your finger shorter. Instead, this body distortion results from the internal logic of our perceptual system rather than conscious reasoning. The perceptual impression that the shell has a back demands that your finger must be shorter, even though you consciously know it is not.

FOOLING YOU INTO SEEING NOTHING

There are lots of magic tricks that necessitate hiding an object or even a person. In the previous chapter, we explored some of the attentional techniques that can make objects invisible. There are, however, other perceptual phenomena that can create very compelling illusions of seeing empty spaces, thus creating the perfect hiding place for things the magician does not want you to see. The simplest way to hide an object is to obstruct our view of it, for example, by hiding it under a cloth or behind a barrier. In these instances, you might not be able to see the object, but you can still imagine that it is actually there. There are, however, psychological principles that allow you to hide an object so that your mind is fooled into seeing nothing.[15]

Let me explain this principle using a rather racy illusion that captures people's imagination on the internet. The illusion is known as amodal nudity, a psychological principle that allows you to "magically" undress clothed men and women (figure 5.7).

Various people on the internet have tried to convince viewers that the effect results from people's dirty minds and our obsession with sex. However, this is not the case, as Ekroll and his colleagues have shown that the illusion works just as well with a boring, cluttered table (figure 5.8).[16] This illusion explains a very important principle in magic. As you will notice, it is very difficult to imagine that there are objects hidden

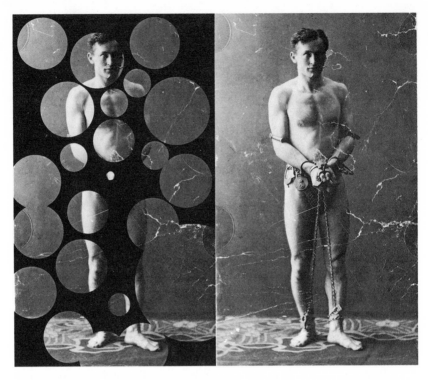

Figure 5.7

Amodal nudity illusion with Harry Houdini

behind the "bubbled" occluders, and Ekroll came up with a very interesting observation. Look at the unoccluded picture and close your eyes, and try to imagine the clutter on the desk. This is fairly easy. Now repeat the same experiment, but rather than closing your eyes, look at the occluded version of the picture. It is much harder to imagine the desk with its clutter when you have the occluded picture in front of your eyes. This illustrates that seeing the occluders interferes with your ability to imagine things behind them, even though you are consciously aware that they are there. Ekroll has suggested that the effect must rely on active suppression rather than a simple failure to represent invisible things. In other words, it is not that you are not seeing the objects but rather that you are seeing an empty space.

This perceptual illusion plays an important role in many magic tricks. For example, the Downs Palm is a palming technique commonly used in coin magic that allows the magician to show his hand empty and then

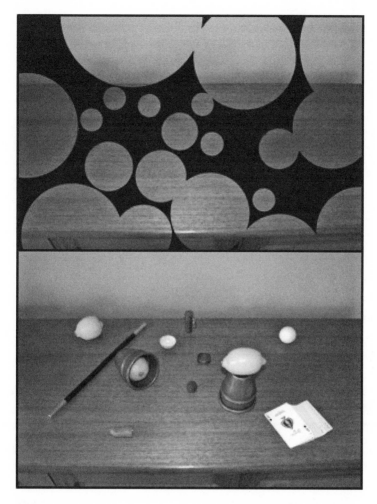

Figure 5.8

Amodal absence illusion

proceed to pluck a coin out of thin air. The secret involves hiding the coin behind the thumb. If you look at figure 5.9, you will experience the hand as being empty. What makes this trick so compelling is that, as you look at the magician's hand, your brain simply can't help but see an empty space.

Ekroll has a simple explanation for this amodal absence effect, and its secret lies in the clever tricks our brain uses to make sense of the world. We have previously learned that perception involves lots of interpretations

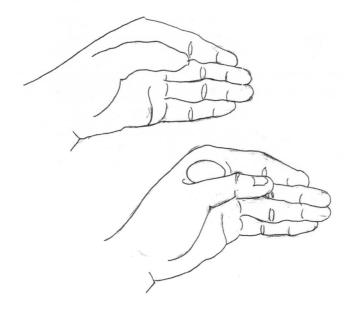

Figure 5.9

Downs Palm: If viewed from the side (top), the hand looks empty even though a coin is concealed behind the thumb (bottom).

and guesswork. It makes complete sense that we should avoid interpretations that are very rare, such as an unlikely coincidental alignment of objects. In the case of the cluttered desk, I spent a long time planning which parts of the picture I should reveal, and a simple misplacement of the occluders would have uncovered some of the objects. A perceptual interpretation that there are lots of objects behind the occluders is unlikely to happen by chance, and it is much more likely that this perceptual image resulted from an empty desk. Seeing truly is believing.

The illusions we have looked at thus far occur because our brain tries to make sense of the world around us. We try to reconstruct the third dimension from two-dimensional images and try to segregate objects from their background and generally organize the visual world in a meaningful way. Scientific studies into these principles are revealing that they are more compelling than most of us had imagined and, in the hands of a magician, can be used to create apparent miracles. We will now look at a different group of illusions that deal with one of our most fundamental perceptual mechanisms: our ability to perceive and interact with a dynamic world.

Understanding the tricks that your brain plays on you will completely transform the way you think about perception. What if I told you that you can see the future? During my PhD studies, Romi Nijhawan told me about the concept of future seeing, and I thought he was completely crazy. I therefore do not blame you for having similar doubts when I tell you that you are indeed able to perceive the future. The principle of future seeing is commonly exploited by magicians but not necessarily in the way that first comes to mind.

SEEING THE FUTURE

Norman Triplett was a pioneer in the psychology of magic, and back in 1900, he published a wonderful scientific paper on magic that, among many other things, discusses an experiment on an intriguing magical illusion.[17] A magician sat at a table in front of a group of schoolchildren and threw a ball up in the air a few times. Before the final throw, his hand secretly went under the table, letting the ball fall onto his lap, after which he proceeded to throw an imaginary ball up in the air. Described like this, it does not sound like an amazing trick, but what was truly surprising is that more than half of the children claimed to have seen an illusory ball—what Triplett referred to as a "ghost ball"—leave the magician's hand and disappear somewhere midway between the magician and the ceiling. This was clearly an illusion because on the final throw, no ball had left his hand; the children had perceived an event that never took place.

Triplett carried out several studies using this illusion, and he came to some rather interesting, though not necessarily correct, conclusions. He thought that the illusion resulted from retinal *afterimages*, or in his own words, "What the audience sees is an image of repetition, which is undoubtedly partly the effect of a residual stimulation in the eye, partly a central excitation."[18] At the time, this seemed to be a reasonable suggestion. I came across Triplett's paper in my early days of researching scientific studies on magic, and I was intrigued by this illusion. Triplett's Vanishing Ball Illusion relies on a principle that I often used to vanish objects, so I had some ideas as to why the illusion worked. I was skeptical about Triplett's explanation, and I knew from experience that the illusion relies on misdirecting the audience's expectations so that they anticipate you throwing the ball for real. A person's eye gaze provides one of the most powerful tools to misdirect expectations, and so I embarked on one of my first scientific projects to study the role that social cues play in driving this illusion.[19]

I recorded two different versions of the Vanishing Ball Illusion. In the normal version, I threw the ball up in the air twice, before secretly palming it in my hand and simply pretending to throw it up in the air (see figure 5.10). Importantly my gaze followed the ball, and in the pretend throw, my gaze followed the imaginary trajectory of the ball. In a different version of the trick, I carried out exactly the same actions, but this time, my gaze did not follow the imaginary ball. Instead, I stared at the hand that was palming the ball. Triplett's account of retinal afterimages predicts that both versions of the trick should be equally effective, because both groups of participants observed exactly the same sequence of events and thus had the same sensory stimulation.

Doing science in the twenty-first century gave me the advantage of new scientific instruments. I could measure participants' eye movements while they watched video clips of the trick, in order to gain insights into why they experienced the illusion. The findings from this study were surprising: In the normal version of the trick, nearly two-thirds of our adult participants experienced the illusion and claimed that they had seen the magician throw the ball up in the air and that it had left the screen at the top. When I asked them how they thought the trick had been created, these participants typically claimed that someone must have caught the ball at the top or that it had stuck to the ceiling. The illusion does not work for me because I know how it is done; indeed, once you know the secret, it will not work for you either. However, most of the participants in the experiment were fully convinced that they had seen the ball move up and were truly astonished when they watched the video again.

Figure 5.10
Vanishing Ball Illusion

My intuition about the role of my gaze was correct, for the illusion was far less effective when I looked at my hand that was concealing the ball. These findings reveal some interesting insights into the illusion. They illustrate that the illusion is mostly driven by expectations, rather than perceptual afterimages. More recently, we have shown that even when you simply pretend to throw a ball up in the air without ever having thrown the ball for real, more than a third of people still experience the illusion.[20] In some ways, we are behaving like dogs who run after the stick their owner simply pretends to throw.

The eye-tracking data revealed some very interesting insights into why the illusion works. After participants watched the video, we asked them about where they were looking, and most of them claimed that they were simply looking at the ball. This, however, was not the case: although they looked at the ball once it had been thrown in the air, they spent a lot of time looking at my face, particularly before the ball reached the top of the screen. What this shows is that participants used my face to predict where I would throw the ball, which turns out to be a very clever strategy.

We live in a dynamic world where things change rapidly, and we often need to react very quickly. For example, when driving, it is important to anticipate what others are doing so that you can react quickly. One of the main differences between novice drivers and experts lies in their ability to predict what others will do in the future, and eye-tracking studies have shown that as you become more proficient, you start to look further ahead.[21] I was one of the first people in the world to learn to drive wearing an eye tracker, and this data revealed that the first time I sat behind the wheel, my eyes were pretty much stuck to the hood of the car. However, with experience, I managed to look farther ahead and to use cues in the environment to anticipate future events.

Watching me toss a ball in the air is far less complex than driving, but perceiving a fast-moving ball is much more difficult than one would expect. Seeing a small ball requires us to look at it, but our eyes are bad at tracking fast-moving objects. Seeing the ball requires your eyes to be looking at the right place at the right time, and you have a very short time window to plan an eye movement to the correct location. Looking at the ball is only possible if you can predict where it will be in the future, and this is exactly what our participants did. In other words, they were using my gaze to predict where I would throw the ball.

The eye movement data revealed another interesting finding: although most of our participants experienced an illusory event, the eyes were not tricked. When the ball was thrown for real, most of the participants managed to look at it when it reached the top of the screen. During the fake throw, participants claimed to have seen the illusory ball at the top of the screen, but they did not move their eyes there, which suggests that our eyes are resilient to the illusion. This result took us by surprise, but it dovetails with several other findings and highlights another truly amazing feature of visual illusions.

We typically think about vision simply in terms of seeing things, but most of the time, we use our vision without consciously realizing that we do so. For example, as you reach for your cup of coffee, you use visual information to guide your arm movement and adjust your grip so that you can successfully pick up the cup. Back in 1995, David Milner and Melvyn Goodale proposed an influential theory of perception that suggests that the visual processes we require for perceiving objects are driven by separate visual pathways than those involved in visually guided actions, such as picking up an object.[22] Some of the most compelling evidence for this theory comes from a neurological patient who goes by the name "D. F." D. F. sustained severe damage to the ventral part of her brain, the pathway that is responsible for object perception. She now suffers from a special condition known as agnosia, which means that even though her vision is intact, she fails to recognize objects and simple shapes and can't even distinguish between horizontal and vertical lines.

One would imagine that not being able to recognize even the simplest of shapes would severely limit her day-to-day capabilities, yet when I met D. F. for the first time, I was astonished by how normal her behavior seemed. She was perfectly capable of navigating a cluttered room, and during lunch, she managed to pick up her knife and fork without problems. How is this possible? D. F.'s lesion has impaired parts of the brain that are responsible for object perception, while the neural systems that are responsible for visually guiding her actions are still intact. As these two systems are anatomically and functionally independent, D. F. is perfectly able to use her vision to guide her actions, while being consciously unaware of the objects around her. For example, even though she is incapable of consciously distinguishing between different line orientations, she is perfectly capable of picking up a letter and putting it in a mailbox.

Picking up objects and interacting with our physical world requires accurate representations of the objects around us, and the perceptual errors that we have discussed so far imply that we should constantly misjudge our actions. Although my perceptual system is often fooled by illusions, I rarely misjudge where to reach. The reason why I don't make these mistakes is that the part of the visual system responsible for orchestrating these actions is not fooled by visual illusions. For example, Richard Gregory and colleagues have shown that if participants are presented with a hollow mask and are asked to point to the nose, they point to a location outside the mask.[23] This is because they consciously perceive the face as being solid. However, if you ask them to quickly flick the nose, their hand moves inside the mask and touches the correct location. This is because flicking is a visually guided action that is driven by the dorsal stream, the visual system that requires reliable spatial information, and thus is not fooled by the illusion. Our eyes are also driven by the dorsal stream, so even though your conscious perception has been fooled by the illusion, your eyes have not.

Our visual system has evolved so that we can interact with the world around us, but why are we so easily tricked into seeing an illusory ball? The answer to this question lies in one of the most surprising and possibly unsettling features of our perceptual systems.

My morning commute takes me through some of London's busiest transportation hubs, yet I am perfectly capable of navigating my way through the crowds and rarely collide with fellow commuters. One of my other truly astonishing skills involves throwing a ball one meter into the air and catching it. "So what?" you might ask. "Most people can do these things with ease, so why should I be amazed?" These tasks are truly incredible, but a full appreciation requires us to look at the challenges in perceiving our dynamic world.

We see things because objects reflect light that is projected onto our retina, and once our photoreceptors register the light, they send a neural signal down the optic nerve. As we have learned earlier on, perception does not take place in the eyes, and lots of complex neural computations are required before we can experience the world. Neural signals are initiated in the retina and then pass via different neural centers to the visual cortex and higher cortical areas, which eventually build a mental representation of the outside world. Neural processing is not instantaneous because neural signals are passed along neurons at a finite speed. It takes about a tenth of a second for the light registered by the retina to become a visual perception in the brain. This means that our perception lags about a tenth of a second behind what is happening in the world. I will give you a few moments for this thought to settle, and just in case you are still struggling to come to terms with it, let me help you with an analogy: During a thunderstorm, vast amounts of electrical energy are discharged, which results in a flash and a loud bang. As you watch the storm from a distance, you see the lighting before you hear the thunder. This is, of course, because sound travels much slower than light, and so we hear the thunder several seconds after the electrical discharge has occurred. It is the same for perception. The neural delay means that we perceive things at least a tenth of a second after they have occurred.

You might think that a tenth-of-a-second delay makes very little difference to your morning commute, but believe me, this is a substantial delay. Let me put it in context: if you are walking at a modest speed of about one meter per second, a tenth-of-a-second delay will result in you perceiving the world as lagging ten centimeters behind you. This is quite hard to believe because you simply do not experience the world as lagging, and such a perceptual error should certainly result in many early-morning collisions. Likewise, this perceptual delay should make it impossible for you to catch a ball, especially because this perceptual delay does not account for the substantially longer amount of time your brain requires to plan and initiate a motor response capable of catching the ball.

It is only once you start thinking about some of the huge day-to-day challenges our visual system constantly faces that the true wonders of the brain start to emerge. Our brain uses a really clever and almost science-fictional trick that prevents us from living in the past: we look into the future.[24] Our visual system is continuously predicting the future, and the world that you are now perceiving is the world that your visual system has predicted to be the present in the past. This idea takes a bit of time to get used to, which is why the first time Romi Nijhawan told me about it, I thought he must be crazy. However, unless we predict the future, we will always experience the past.

There is a lot of experimental work that supports this idea, and the most striking demonstration comes from a phenomenon known as the Flash-Lag Effect.[25] If you move a ball at a continuous speed and suddenly flash a light just as the ball moves past it, observers perceive the flash as lagging behind the ball. At the time of the flash, the ball and the flash were at the same location, yet they are perceived as being at different spatial locations. As the ball is moving at a continuous velocity, it is easy to predict its location based on clues picked up in the past, and thus we perceive the ball based on where we expect it to be in the future. This is an example of our future-seeing capabilities. The flash, on the other hand, was unexpected and impossible to predict. This means that your perception of the flash was delayed by approximately a tenth of a second, and by the time you experience the flash, the ball has already passed. Although these two events occurred at the same location at the same time, your failure to predict the flash resulted in a delay in perception, and thus you experience them as being spatially apart.

There are lots of other experimental demonstrations that illustrate how we perceive and remember events not necessarily as they are now but as we expect them to be in the future. For example, if you are asked to remember the location of a moving object, you will remember the object farther along its movement trajectory, a phenomenon known as representational momentum.[26]

We have learned much about perception since Triplett's time, and we now believe that people experience the Vanishing Ball Illusion because they perceive events based on what they expect to happen in the future. Indeed, our more detailed analysis of the Vanishing Ball Illusion reveals that it likely results from a perceptual system that predicts at least a tenth of a second into the future—a system that sees the present.[27]

There are lots of other magic tricks that exploit the brain's predictive powers. For example, Jie Cui and colleagues have shown that when master magician Mac King pretends to toss a coin from one hand to the other, people perceive the coin as flying between then.[28] Unlike the Vanishing Ball Illusion, which only works once, this trick works even when you know how it is done; you can repeat it over and over again. More recently, Matt Tompkins and colleagues have studied an intriguing magic trick that shows how people can perceive objects that they expect to be present.[29] In the Phantom Vanish, the magician reaches into his pocket to grab a handful of coins. He then simply pretends to take a coin and uses "sleight of hand" to make the imaginary coin disappear. Even though the coin's presence is simply implied by the actions and never seen, nearly a third of the participants claim to have seen the coin. Most astonishingly though, people are genuinely surprised when the imaginary coin vanishes; they are surprised when nothing turns into nothing.

The perceptual distortions we experience in magic provide intriguing and often surprising insights into the nature of perception. Although some magic tricks are based on well-understood perceptual principles, others involve lesser-known processes. All of these tricks push our perceptual system to its limits, and they offer scientists new test beds for studying the brain. For example, we have used the Vanishing Ball Illusion to study perceptual and attentional abnormalities in autism. Previous research suggested that individuals with autism focus on local sensory detail, which makes them less susceptible to typical visual illusions.[30] However, at the time, very little was known about how they would respond to more complex illusions that exploit predictive vision. Individuals with autism are also less sensitive to social cues, and given that the Vanishing Ball Illusion relies heavily on social cues, we predicted that they would be less susceptible toward the illusion.[31] We were therefore surprised to find the opposite pattern of results: our highly functioning adults with autism were actually more sensitive toward the illusion.[32]

As scientists, we are usually disappointed when our experiments don't turn out as predicted, but sometimes, these surprising results can provide even more interesting insights. As we looked at our eye-tracking data more closely, we discovered that the individuals with autism were far less efficient at using the magician's social cues to predict the ball's trajectory, which meant that they failed to see the ball when it was thrown for real. Because they did not have an accurate reference point to which they could compare the fake throw, it was very difficult for them to distinguish between the real throw and the implied throw. Hence, they were more likely to see a ball when it was not there. Our ability to distinguish between illusion and reality depends on both the strength of the illusion and our knowledge of what reality truly looks like.

Our visual system has evolved so that we can interact with our world, and many of the everyday perceptual tasks that we take for granted require vastly complex computations. Other animals, such as birds of prey, have more acute vision, enabling them to see four to five time farther than we can.[33] Although human beings are not necessarily the king of acuity, our visual processing abilities are unrivalled by any other species. For example, we are extremely good at recognizing a vast array of different objects, even under extremely difficult conditions, and we are exceptionally good at quickly adapting our behavior to a continuously changing environment. In this chapter, we have seen that our visual experience is a product of a vast range of computations. Understanding some of the complexities of the task at hand helps us appreciate that our perceptual experience is more subjective than we intuitively think. Seeing David Copperfield's cheeky smile, despite the fact that his body has been visibly cut in half, should remind us that we can't necessarily trust everything we see. Perception is the first step in a hierarchy of cognitive processes, and as we will see next, even the things we correctly see can be distorted and twisted in our memory. However, for the moment, let us simply reflect on the fact that even though we think that we believe in the things we see, we actually see the things we believe.

$\textcircled{6}$

MEMORY ILLUSIONS

THE ASIAN MAGICIAN STOOD in front of a wicker basket, and after a few magical gestures, a rope mysteriously snaked up into the air and remained erect. His boy assistant then scurried up the rope and, once he reached the top, mysteriously vanished. The magician called for the boy to come back, but he refused to return. The magician became annoyed, climbed up the rope himself, and also vanished. Moments later, dismembered parts of the boy's body fell to the ground. Covered in blood, the magician descended the rope and placed the body parts into the basket. A few magical incantations later, the boy jumped from the basket, fully restored to life. The classic version of the trick was performed during the day, in the open, and with the magician completely surrounded by observers.[1] The Indian Rope Trick has captivated magicians and the general public for many years. The main reason for this fascination is that the illusion is entirely impossible.

The exact origin of this miracle is unknown, and whilst numerous explanations were offered, [2] these reports generally lack detail, and full descriptions of the trick did not arrive in the West until 1890. That year, the *Chicago Daily Tribune* printed a story describing how an "American visitor had witnessed an Indian juggler performing a trick in which a rope rose in the air, a boy vanished at the top, his limbs fell to the ground, and was restored to life."[3] At the time, it was thought that the trick was produced through mass hypnosis, and the story created much interest and was republished in several foreign newspapers. It turns out that the original story was a hoax, and although the paper published a retraction, this received little attention, and the myth of the Indian Rope Trick was born. However, in spite of the original story being a hoax, numerous eye-witnesses appeared who all claimed to have seen the miracle with their own eyes.

In 1934, the Occult Committee of the Magic Circle held a meeting to investigate the evidence for the trick and came to the conclusion that the Indian Rope Trick was a myth. The committee was so convinced by its verdict that it offered a substantial reward to anyone who could perform the trick under the original conditions. The reward remains unclaimed. Far from quelling the myth, however, each investigation has brought forth further eyewitness accounts. Although many magicians have performed similar versions, they have never managed to replicate the original effect. They typically rely on clever stagecraft, but the original trick was said to have been performed in broad daylight, in the open, fully surrounded by people. As it stands, the trick simply cannot be performed by methods known to either magic or science.

Inasmuch as it is impossible to travel back in time, we cannot verify or refute these eyewitness accounts. However, it is very unlikely that the Indian Rope Trick was ever performed in the way it is described. But why did people claim to have seen it? Were they lying?

It is possible that the eyewitnesses genuinely remembered seeing the trick as described. One of the biggest joys of being a magician involves listening to other people's accounts of your own performance. Spectators often exaggerate what actually happened, miss critical parts, and embellish the performance with colorful descriptions of things that never took place. Back in the late 1880s, Richard Hodgson and S. J. Davey documented that people systematically misremembered key aspects of a staged séance, and these systematic memory distortions explained why the sitters experienced impossible phenomena.[4] In this chapter, we will explore how magicians distort and erase your memory traces, as well as how and why people remember impossible phenomena.

HUMAN MEMORY

Memory is one of the most important cognitive functions, and most of our daily activities rely on storing and retrieving information. Doing so relies on clever tricks and shortcuts that can lead to mistakes. Human memory is far from perfect, and we frequently experience memory failures, such as when you forget someone's name or fail to recollect the first time you met. However, most memory glitches go unnoticed, which might explain our inflated confidence in our ability to recall past events.

Christopher Chabris and Daniel Simons report that more than 60 percent of the population believes that our memory works like a video camera, accurately recording the events we see and hear so that we can review and inspect them later.[5] Although the video camera analogy is intrinsically appealing, it fails to capture the true nature of human memory.

Most of our memories start with a perceptual experience, and these experiences are highly subjective. Our senses are bombarded with vast amounts of sensory data, and we only select the information that is relevant to us. We can only truthfully remember things we actually perceive, and because our perception is more limited than our experience suggests, we can only assume our memories are incomplete. Our perceptual system is amazing, but it is far from perfect; it involves lots of interpretation and guesswork. The first step in forming a memory trace involves encoding the incoming information, and our memory is only ever as reliable and complete as the perceptual experience it is built on. As we have seen in previous chapters, this foundation is much more wobbly than we naturally assume. It should therefore come as no surprise that our memories are certainly not perfect.

Human memory is immensely complex, and we store different types of information in different systems. For example, we use short-term memory to store small amounts of information for brief durations (e.g., telephone numbers). Anyone who has attempted to remember a telephone number will appreciate that our short-term memory has a very limited capacity— about five to nine pieces of information. Short-term memory acts as a processing space, which is why it is commonly referred to as working memory.

Only a fraction of the information we hold in short-term memory ever enters the long-term store, and our brain forms different types of long-term memories. For example, we store unconscious procedural memories that allow us to perform tasks such as writing or riding a bicycle. We also hold general knowledge in the form of semantic memories (e.g., London is the capital of the United Kingdom), but we rarely remember the time when these memories are formed. Details about specific events are stored as episodic memories. Understanding why people misremember a magic trick requires a close look at these episodic memories.

Memories are stored in complex neural networks, and it is tempting to think of the brain as a large computer. Parts of this analogy are correct. For example, computers and brains both have different components that deal with short- and long-term memories. Computers use RAM, which has a relatively small capacity but processes information quickly. However, most of the information is stored on larger, more sluggish hard drives. Similarly, the brain stores different types of memories in different anatomical structures, all of which have different properties. There are, however, some fundamental differences between human and computer memory. First, our memories are stored in distributed networks, rather than discrete physical locations. Second, rather than simply being deposited deep inside our brain, new information is continuously consolidated and connected with information that we already have available to us. Unlike computer memory, which lies dormant until it is accessed, human memory is continuously updated and modified.

Data is only useful if you can access it, and retrieving a relevant piece of information is not simple. Computers build huge databases that index where the relevant information is stored. Without such indices, the information itself is futile. Retrieving our own memories is far from trivial, and there are countless factors that influence the ease with which memories are retrieved. For example, your emotional state and even your physical surroundings determine which past memories come to mind.[6] You remember more positive instances when you are happy than when you are sad, and you are more likely to recall things when you are in the same physical location as where the memory traces were formed in the first place. Just because you have a memory trace does not mean that you can remember the event, and there are lots of psychological factors that influence how your memories are retrieved.

As we think about the events we have experienced, we are only ever aware of the content that we recall. Remembering an event involves an interactive cascade of cognitive processes (encoding, storage, retrieval), and errors can creep in at any one of these stages. As you reflect on instances where you have forgotten or misremembered an event, it is impossible to know where the problem occurred. Did you fail to remember an event because you simply did not encode it? Maybe the information was encoded but faded away over time. Alternatively, the memory trace might be stored deep in your brain, but you simply fail to access it at this particular point in time. Identifying the true source of the error is challenging, but scientists are starting to get a grasp on some of the factors that influence how we remember the past.

FORGETTING

Juan Tamariz was the guest of honor at the 2017 Science of Magic Association Conference, and those of us in attendance had the pleasure of experiencing a true master of memory misdirection. We sat around a small table watching Tamariz create miracle after miracle. In one instance, he instructed me to keep my eyes on a pack of cards that was placed on the table in front of us, after which he asked one of us to simply name a card. Even though he never touched the cards, this freely chosen card (and I am pretty convinced it was not forced) was the only card missing from the pack. We were all perplexed, and I have absolutely no idea as to how this was done. I had experienced something truly impossible.

My description of Tamariz's performance is as implausible as the eyewitness accounts of the Indian Rope Trick. Unless Tamariz has genuine supernatural powers, which he adamantly denies, I can only assume that my recollection of the trick must be false. Tamariz is a master of memory misdirection, and my only rational explanation is that he was playing with our minds.

Memory misdirection is an important technique that allows magicians to erase memories that contain crucial information about how a magic trick works. Let me explain memory misdirection using one of Tamariz's own examples.[7] The magician gives a spectator a deck of cards, which she is required to shuffle. He then hands her an envelope and asks her to examine it thoroughly, after which he gives her a pencil to sign it. While she signs the envelope, he offers some help and takes back the deck of cards, after which he immediately returns the cards and takes two steps back. He now asks her to take the top card from the deck and place it inside the envelope. Now the magician engages in some elaborate patter to flesh out the trick, after which he reveals that the apparently freely chosen card inside the envelope matches a prediction that he made before the trick began.

The method for this effect is simple: the magician secretly adds a card to the top of the deck after the spectator has shuffled the cards. The helpful gesture of holding the cards while the spectator signs the envelope offers a perfect moment to do so. This trick does not rely on perceptual misdirection, because there is no need to mask the action of adding the card to the deck. This all happens in full view. The key to this trick lies in ensuring that people forget that the magician touched the cards after they were shuffled. At the end of the trick, people need to believe that the selection of the top card was based on pure chance. Erasing the memory of the magician touching the cards ensures that the effect is impenetrable.

How can you erase a person's memory? Tamariz explains that there are two stages to this process. First, the magician must let some time elapse between the method and the effect. As time passes, our memories start to fade, and time is one of the most powerful psychological processes that lead to forgetting. This is why Tamariz embellishes the performance with elaborate antics, which might include hanging the envelope from a thread to ensure it cannot be tampered with. All of this is irrelevant to the trick, but it acts as a natural way of delaying the effect and bombarding you with new experiences that help displace your memory of seeing him touch the cards. Second, Tamariz employs a verbal suggestion that helps erase that particular memory. In an attempt to help jog your memory, he recounts the most important parts of the trick: "You examined the envelope, shuffled the deck, and placed a card inside the envelope, which you signed yourself. This means that the card was freely chosen, without the possibility of it being switched by any manipulation on my part, since I have been distanced at all times from it." Most of this information is correct, but he fails to mention that he touched the cards. According to Tamariz, the spectators will now forget that he held the cards for a few seconds, and this simple memory extinction turns a normal sequence of events into an impossible miracle.

I often use this technique in my own performances, and I am truly astonished by the ease with which people's memories can be erased. However, based on what we have learned so far, these types of memory failures should not be surprising. Change blindness highlights astonishing short-term memory failures, and it demonstrates two surprising limitations. First, unless we attend to things in our surroundings, we simply do not perceive them—let alone remember them.[8] Second, people vastly overestimate their ability to notice things, and they misjudge the amount they remember.[9] These bewildering short-term memory failures raise serious questions about the amount that we are able remember a few minutes, hours, or days after an event.

We remember far less than we think we do, and I'm happy to illustrate this point using a simple example. Pick up a coin, look at it, and keep it in your closed hand. Most coins have a portrait on one side, but which way is the person looking? I am sure you have handled this type of coin many times and have spent much time looking at it, yet you would probably struggle to draw the coin from memory. Raymond Nickerson and Marilyn Adams have shown that people barely remember the visual details of a common object.[10] Volunteers were asked to draw a US penny and to choose the features that appeared on the coin from a list of possible descriptions. Their results were conclusive: people simply did not remember this type of detail.

Our memory does not work like a video camera, capturing and storing every single detail we experience. Storing information requires cognitive resources, and it simply does not make sense to waste resources on unnecessary information. For one, data storage requires physical hardware, and our neurons and their connections require energy and space. More importantly, the more information we store, the harder it is to retrieve it efficiently. Storing every single sensory experience would clog up our brains with unnecessary clutter. Our sensation of remembering detailed, truthful memories is just as much an illusion as the grand illusion of perception. But why do we fail to notice these memory illusions?

We do not notice the huge gaps in our conscious experience because the grand illusion is rarely challenged. The same is true of memory. It is easy to think of instances when we have forgotten something, and we often notice these memory failures. But what about the times when we believe that we are truly remembering an event? Does our conscious recollection prove that we are genuinely remembering it? The intuitive answer is yes; much of our legal system treats assertive eyewitness testimonies as reliable evidence. However, just because we are confident in remembering something does not mean that we have experienced it.

We treasure our own memories, and questioning someone else's is akin to questioning their integrity. Nobody likes to be accused of misremembering, let alone of making things up. But what if I told you that most of what you remember is simply a reconstructed approximation? As you remember an experience, you are not remembering the event itself but your own interpretation of what may have taken place. Your memory of the past is an illusion because the memory itself is simply a reconstruction of the event.

You rarely notice these memory illusions because you are tricked into believing that you are remembering the actual event. For example, people often claim to have vivid memories of events that are of personal significance. Back in the 1970s, Roger Brown and James Kulik coined the term "flashbulb memories" to describe people's detailed memories of emotive events.[11] At the time, most people confidently remembered the moment when they first heard that John F. Kennedy had been shot, and many subsequent studies have reported similar findings following other national tragedies. Arousal enhances people's memories, but more recent studies question the accuracy of these flashbulb memories. For example, Kathy Pezdek tested people's memory seven weeks after the tragic events of September 11, 2001.[12] Even though most respondents confidently remembered the day in detail, some of these memories may have been false. For example, 73 percent remembered seeing video footage of the first plane striking the North Tower of the World Trade Center. George W. Bush also recalled that he first found out about the attack when he saw the TV images of an airplane crashing into the first tower. This was not possible, however, because the videotape of the first plane hitting the Twin Towers was broadcast the day after 9/11. Conspiracy theorists often cite this as conclusive evidence that Bush must have known about the attack before it happened. However, it is much more likely that Bush, like most of us, simply misremembered the tragedy. The video footage of the first plane crashing into the North Tower has become one of the most harrowing and iconic images of the early twenty-first century, and this video clip has been played on televisions all over the world, rewriting our memories. Although we feel as if we have detailed memories of these emotional events, they are no more accurate than our memories of more mundane occurances.[13] The vivid details we seem to remember add to the conviction that a memory must be true. This might, however, simply be an illusion.

MISREMEMBERING

Back in the 1930s, Frederic Bartlett observed that people do not simply remember specific instances and instead recover information about general themes and fill in information that is consistent with these themes.[14] Bartlett asked his students to read a disjointed Native American story entitled, "War of the Ghosts," after which they had to recall it in their own words. Although the original story did not make much sense, his students reinterpreted it in light of their own knowledge of the world and imposed order on this rather muddled story. By the time they recalled the story, it had started to make more sense. Bartlett suggested that memory is a reconstructive process that is largely driven by our own knowledge and experience. Bartlett's observations were important because they demonstrate that memory is not a faithful recording or readout of the events we have experience; it is malleable. Our memories are highly subjective, and there are lots of psychological factors that influence how we remember the past.

Henry Roediger developed a simple paradigm that shows that people often remember things that are implied but not necessarily stated.[15]

Participants were asked to remember a list of related words, such as "bed," "rest," "awake," "tired," "dream," "wake," "snooze," "blanket," "doze," and "slumber." Later on, they were given a recognition test that included the original words, as well as some new words that were semantically related, such as "sleep." Although the word "sleep" had not been included in the original list, participants frequently claimed to have seen it. They falsely remembered an event simply because it related to an event they had experienced.

If false memories are indistinguishable from real past experiences, we should be able to change our past. It is indeed relatively easy to implant false memories after an event has occurred. Elizabeth Loftus and her colleagues have spent the past four decades exploring ways in which you can change people's memories.[16] In their seminal study, they had participants watch a video depicting a traffic accident and then asked them questions about how fast the cars had been traveling when the accident occurred. Crucially, however, the researchers used two different verbs to describe the accident. Participants who were asked "How fast were the cars going when they smashed each other?" estimated the speed at forty-one miles per hour. However, when the experimenters used the word "hit" to describe to collision (instead of "smashed"), the estimates dropped to thirty-four miles per hour. Even more surprisingly, the wording of the question distorted people's memories. A week later, the participants were asked whether they had seen any broken glass in the film. Thirty-two percent of those who had received the leading question with the word "smashed" mistakenly reported having seen broken glass, while only 14 percent of the people in the other group made this error. A single word in a single question was sufficient to alter people's memory of an event.

Over the past few decades, several different ways of planting false memories have been discovered. For example, one study enlisted the help of family members to persuade participants that they had been lost in a shopping mall during childhood, before being rescued by an elderly person and reunited with their family. About a quarter of the participants later falsely claimed to remember this fabricated story.[17] It is possible that such an event did take place, and thus it is difficult to prove that the memory was false. However, suggestive memory techniques have been used to implant memories that are simply impossible. For example, volunteers were asked to evaluate a fake advertising brochure describing a visit to Disneyland, which described how they met and shook hands with Bugs Bunny. After reading this fabricated brochure, 16 percent of the participants claimed that they remembered shaking hands with Bugs Bunny.[18] Given that Bugs Bunny is a Warner Brothers creation, we can be sure that this event never took place. Nevertheless, people frequently reported precisely what they remembered about their fictional encounter: 62 percent remembered shaking his hand, 46 percent remembered hugging him, and some remembered touching his ears or tail.[19]

There are other powerful ways in which false memories can be planted into our minds. For example, Kimberley Wade and colleagues used doctored photographs depicting the volunteer as a child on a hot-air balloon ride.[20] The volunteers were presented with the fake photograph and were asked to describe everything they remembered. After a few follow-up interviews, half of the volunteers recalled the fictitious balloon ride, and some even embellished their reports with details. For example, one subject said, "I'm still pretty certain it occurred when I was in sixth grade at, um, the local school there. ... I'm pretty certain that mum is down on the ground taking a photo." These vivid memories must have been false because none of them had ever flown in a hot-air balloon as a child.

One simple way of changing a person's memory is to manipulate the context in which the information is encoded. Richard Wiseman and Emma Greening showed volunteers a video clip of a psychic using magical powers to deform a key.[21] He then placed the bent key on the table, and half of the participants were given verbal suggestions that the key was continuing to bend. The other half of the participants were not given any further suggestions. The participants who received the suggestions often later reported that the key continued to bend once it was on the table, and they even forgot about the verbal suggestion.

Magicians often use these types of memory suggestions. The Vanishing Ball Illusion is another powerful example of how people remember events that are implied but have never taken place. In the Vanishing Ball Illusion, people remember seeing a ball leave the magician's hand after which it disappears in the air.[22] This recollection is very different from what actually takes place; the magician simply pretends to toss the ball in the air. We currently don't know whether this illusion relies on people misperceiving the event or whether they simply misremember it, but it likely involves both. Manipulating your expectations about what you believe is happening will strongly influence how you remember the event in the future.

Another way of implanting false memories is to ask people to simply imagine that an event has taken place. Mental imagery is a powerful tool to enhance your memory, and some mnemonic strategies rely on forming visual images of the items to be remembered. These techniques are surprisingly effective, but they can also distort your memories. For example, simply imagining a fictitious childhood event can result in falsely remembering it.[23] More dramatically, Ayanna Thomas and Elizabeth Loftus asked volunteers to either imagine or carry out bizarre actions such as kissing a mirror or sitting on dice. A few weeks, later many of the participants falsely remembered doing the actions that they had only imagined. This effect was even stronger for more usual actions, such as flipping a coin.[24]

Many of these memory distortions result from *source-monitoring errors*, in which people incorrectly attribute recollected experiences. For example, you might hear about a story from your friend and later report having read about it in the newspaper. George W. Bush had a vivid memory of seeing the first plane crash into the World Trade Center and falsely attributed this memory to first hearing about the attack. Hillary Clinton made a similar attribution error in 2008, when she claimed to remember landing under sniper fire when arriving in Bosnia, a memory distortion exploited by her political rivals. Source monitoring errors are extremely common, and magicians frequently use them. For example, when Tamariz summarizes the sequence of events in his card trick, he will purposefully omit certain details (e.g., that he touched the cards) and add further false information (e.g., "You shuffled the cards"). These verbal statements automatically activate mental representations of the events. Later, when you try to work out how the trick was done, you remember something about shuffling the cards, but you might fail to recall the true source of the information. Hence, you falsely believe that you actually carried out the action.

We rarely watch magic on our own; in most instances, it is performed for a group of people. After the show, people enjoy discussing the trick with their friends, and they try to work out how it was done. Group work can add new perspectives and potentially facilitate the discovery of the method, but it can also lead to errors. Humans are highly social, and collaborating with others has given us a huge adaptive advantage. Successful collaboration often necessitates compromises, but what happens when your collaborators make mistakes?

Solomon Asch conducted seminal studies investigating how people respond when they are given conflicting information.[25] In these experiments, a volunteer was told that he was participating in a visual acuity test in which he had to report which of three lines matched a reference line. The task was very simple; when tested alone, people rarely made any mistakes. The visual acuity test was a cover story, as Asch was interested in determining whether people would yield to group pressure. The volunteers were tested in small groups (seven to nine people), and each person had to report their responses aloud in the order that they were seated. All but one of the participants were confederates pretending to be subjects, and they were instructed to deliberately give the same wrong answer. When faced with such overwhelming conflicting information, the real subject, who reported second to last, conformed to the erroneous judgments more than a third of the time. These experiments show that social pressure results in people denying the evidence of their own eyes and yielding to group influence.

Social conformity can also influence people's memory of events. Dana Schneider and Michael Watkins asked volunteers to remember a list of words in pairs, and unbeknown to the volunteers, the partner was a confederate.[26] After studying the words, subjects had to verbally report whether or not certain words had been presented. On a number of occasions, the confederate falsely claimed that new words had been presented and rejected genuine items. This misleading information influenced people's recognition performance. They falsely recollected the new words that had been accepted by the confederate and rejected genuine ones that had also been rejected.

Our memories are influenced by social factors, and discussing what you have just witnessed with another person has serious implications for how you remember an event. Fiona Gabbert and colleagues developed a clever paradigm in which two people watched video clips depicting the same event, but the footage was shot from different angles, and thus each video contained unique items that could be seen by only one of the participants.[27] For example, the video showed a girl entering an office to return a borrowed book. In one version, you could read the title of the book and also see her throwing a note into the dustbin. None of this could be seen in the second version. However, in that version, you could see her sliding a £10 note from a wallet and putting it into her pocket. After each participant had watched a different video clip, they were then invited to discuss with each other what they had seen before being questioned in isolation about what they remembered. It is important to note that the participants did not know that they had watched different videos. Rather surprisingly, 71 percent of the participants mistakenly claimed that they had witnessed seeing things that they had heard in the discussion but that did not happen in the video they had watched. For example, many of the participants who could not have seen the girl steal the money now suddenly claimed to have seen her take the note from the wallet.

These findings have serious implications for the value of eyewitness testimonies. The Oklahoma City bombing in 1995 provides a chilling test case. Key evidence came from three eyewitnesses, who all worked in the shop where Timothy McVeigh rented the truck that he used in the bombing. After his arrest, there was an inquiry into whether he had acted alone or had an accomplice. One of the people working in the rental shop initially claimed that McVeigh had been accompanied by a second man, while the two other witnesses gave no description of this accomplice. However, later on, these two witnesses also claimed to remember details of the second person. Several months later, the first witness confessed that he may have been recalling another customer. It is likely that the three men discussed the event and that false descriptions molded their memories.

All our thoughts and beliefs rely on how we remember the past. We treasure some memories like precious objects, and the idea of people manipulating or erasing them makes us feel uncomfortable. However, in this chapter, we have seen that our memories are much more fragile and malleable than we think. We are tricked into believing that our memories represent accurate snapshots of our past. In reality, they are far less complete and reliable than we assume. Magicians often manipulate our memories using a range of memory misdirection techniques, which result in people recalling a rather different sequence of events. It is difficult to distinguish real memories from false memories, and as such, it is highly likely that people genuinely remember seeing an event that seems entirely impossible. Just because we remember it, however, does not mean that we have experienced it, nor does it prove that the event ever occurred in the first place.

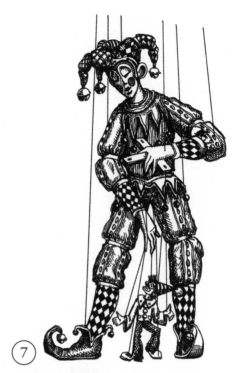

MIND CONTROL
AND THE MAGICIAN'S FORCE

DERREN BROWN IS ONE of the most popular magicians in the United Kingdom. Back in the late 1990s, he pioneered a new type of magic called "mind control," which blends magic and science. Brown claims that his demonstrations rely on a combination of suggestion, psychology, misdirection, and showmanship. While traditional magicians are extremely secretive about their mysterious powers, Brown astonishes his audience by telling them exactly how it is done. He does not pretend to read your mind or to contact mystical spiritual forces. No, Brown claims that all of his illusions rely on manipulating your mind using scientifically plausible principles.

In one such performance, a volunteer named Alice is invited to browse through Hamleys, the United Kingdom's largest toy store, and she is asked to mentally choose one of the nearly quarter of a million toys. After spending a long time deliberating over her choice, Alice finally makes up her mind, but to her absolute amazement, Derren Brown has predicted what seemed like an entirely free choice. Predicting the outcome of a future event is a classic magic trick, but unlike most traditional magicians, Derren Brown explains exactly how this was done. He does not pretend to have time-traveling skills or any other supernatural powers. Instead, he claims to have used unconscious priming techniques that allow him to control Alice's mind. For example, before entering the store, Brown subtly mimed a giraffe and mumbled the word "giraffe" in her ear. As they browsed through the store, they passed by giraffe patterns and giraffe toys, and the word "giraffe" was printed in lots of locations. Brown explains that while Alice did not consciously process these stimuli, they unconsciously influenced her choice. Indeed, when questioned, Alice is completely baffled as to how he managed to predict her free choice.

Derren Brown is one of the most talented and impressive magicians, but are we really that suggestible? In the next two chapters, we will explore how magicians control your mind, and we will take a closer look at the science behind mind control. Is it possible to take charge of your thoughts? What does this mean for your sense of free will in general?

Taken at face value, Derren Brown's demonstration suggests that we are easily manipulated into doing things against our will and that we have very little control over our own thoughts and actions. This will jar with many because we intuitively feel as if we are in charge of our own thoughts and behavior, and the idea of abandoning our sense of free will is unsettling. But is this form of mind control scientifically possible? If so, it raises some fundamental questions about who we are.

The debate about free will dates back to the ancient Greek philosophers, and philosophers and scientists have argued for millennia about whether we are truly free in choosing our actions. Determinism suggests that all our actions have been determined by past events, and this includes the neural processes inside our brain. By its own logic, determinism implies that there is no such thing as free will. Abandoning our free will is deeply unsettling, and doing so has major implications for society. If we are not responsible for our own actions, who is? Can you be convicted of a crime if you are not responsible for what you have done? If all of our actions have been influenced by past events, do we have any choice about being good or bad? Although we find it hard to abandon the notion of free will, arguing for free will raises several serious challenges.

The concept of free will lies at the heart of what makes us human, and answering this issue requires us to look at the fundamental building blocks of the universe. One of Isaac Newton's many revolutionary discoveries was the idea that for every action, there is an equal and opposite reaction. According to Newton's law, all events in the universe are part of a ginormous causal chain that was set in motion at the birth of the universe. Newton's law implies that if you have accurate measurements and a correct model of the universe, you will be able to predict any future event.

Newton's law raises a big challenge for advocates of free will because your every thought and action results from neural activation inside your brain. Your decision to turn the page and all of the fine motor movement required to do so rely on neurons inside your body communicating with one another. Similarly, your decision to continue reading relies on neural activity deep inside your brain. Very few scientists would disagree that all our thoughts and behavior result from complex neuronal activities. Rejecting the biological basis for human cognition involves accepting a form of dualism, but as we have seen in the previous chapter, few accept the idea that our neurons are influenced by some form of metaphysical magical force. However, accepting a biological and physical cause for all of our behavior implies that all of our thoughts are part of the causal chain, which according to Newtonian law implies that they are predetermined. The problem we now face is that unless we accept dualism, our thoughts, including our sense of free will, are part of the causal chain that was set in motion long before we were born.

Abandoning free will is uncomfortable, and philosophers have tried to find ways in which physical determinism and free will can both hold true. This view is known as compatibilism, and physicists and mathematicians have provided new ways in which free will can potentially be regained. During the course of the previous century, it became apparent that Newton's laws were not as universal as he had hoped. For example,

French mathematician and physicist Henri Poincaré discovered that while Newton's laws can explain relatively simple planetary systems, it is practically and theoretically impossible to predict more complex systems.[1] Although these complex systems are determined by a starting point, it is simply impossible to predict their outcome. Our weather represents one of these chaotic systems, and even though you can use a large number of meteorological measurements and complex computational models to predict tomorrow's weather, it is impossible to come up with a reliable long-term forecast.

Edward Lorenz illustrates the nature of these chaotic systems nicely by using a weather metaphor, which is now simply known as the "butterfly effect."[2] These chaotic systems do not behave randomly, and while it is possible to make some fairly accurate short-term predictions, it is impossible to predict the future. However, chaos theory does not necessarily kill determinism. Just because you cannot predict the weather does not imply that it has not been predetermined.

A bigger problem for determinism comes from quantum mechanics. Even though objects that we can observe with the naked eye play by the rules of Newton's laws, subatomic particles play by a very different rule book. In the quantum world, very strange things can happen. For example, quanta can simultaneously act as waves and particles, and on a quantum level, we cannot predict things in a deterministic way. Erwin Schrödinger demonstrated that the simple act of measuring an electron's position as it orbits the atomic nucleus alters its value. This means that, from a theoretical perspective, it is impossible to measure or predict the future. There is a big leap from quantum mechanics and the biological processes underpinning human cognition, yet some have argued that quantum laws can also apply on a cellar level.[3]

In the physical world, other principles, such as that of emergence, propose that complex systems may be more than the sum of their parts, which implies that you cannot necessarily predict their function simply by understanding the theory and laws of another level of organization. For example, understanding the inner workings of the human brain does not necessarily inform us about how humans interact with one another or the abilities and functions that emerge from such interactions. As we see with quantum mechanics, different laws apply to different levels of organization, and it is therefore possible that the laws that govern the complex system underlying our conscious experience cannot be simply reduced to the physical laws of causality.[4] This argument does not explain how our sense of free will emerges, but at least it provides an argument against determinism and thus opens the possibility that we are in charge of our own actions.

The debate as to whether we have free will or not is complex, and more than two thousand years of discussion have failed to result in any conclusion on the matter. One of the central problems lies in the fact that our sense of control is compelling, but some scientists have started to argue that this sense of control might itself be an illusion.[5] Proving that thoughts and behavior can be unconsciously manipulated provides some weight to this argument. However, let's not take Derren Brown's performances at face value, because they too may be illusions. Before we abandon our belief in free will, let us look at the psychology underlying this type of mind control. Are you truly able to control a person's mind through unconscious priming? Can we use hypnosis to create a programmed assassin (as in *The Manchurian Candidate*)? Is our sense of free will really an illusion? Let us start by looking at some of the ways that magicians can influence your mind.

THE MAGICIAN'S FORCE

Imagine that a magician asks you to pick a card from a fully shuffled deck. However, before you do so, he declares that he has made a prediction that is sealed inside an envelope. To everyone's amazement, the prediction matches your freely chosen card. How is this possible? Although your choice may have felt as though it were free, the magician in fact forced you to choose that particular card, a principle that magicians refer to as the *force*. While the force is commonly applied in card magic, magicians also use this principle to influence a wide range of selections. The key purpose of the magician's force is to influence a person's choice without them being aware of it.[6] In some instances, the magician has full control over your decision, while in others he simply increases the probability that you will choose a particular item. It is important to note that forcing is distinct from other forms of social persuasion, such as when a salesperson tries to indirectly yet overtly persuade you to buy his product.[7] In the magician's force, your choice has been systematically biased, but you feel like the selection was entirely free.

There are as many different ways to force your decision as there are to misdirect your attention. It is not my intention to describe them all, nor do I want to reveal the secrets behind some of the cleverest methods. However, let us look at some of the most common forcing principles.

Many forcing techniques rely on restricting your choice by making it physically impossible to choose another item. For example, you may be asked to choose a card from a pack that contains identical playing cards. This is the most basic card force, but there are countless craftier techniques that rely on the same principle. For example, in the classic force, the magician spreads the cards in a particular way and times his spreading action so that your hand reaches for the intended card precisely at the right moment. Although you feel as though you had the opportunity to pick any card, you end up with the card that the magician pushed between your fingers. While the magician forced you to pick that card, you feel like it was your own choice.[8]

Other forcing techniques rely on exploiting people's stereotypical behavior. For example, if you place four cards on the table and ask the spectator to touch one, he is unlikely to touch the cards on the outside and is most likely to go for the one just right of center. Similarly, when you ask someone to choose a number between one and ten, the most common answer is seven. Jay Olson and colleagues have recently measured the probability of naming different paying cards, and their analysis reveals that some cards, such as the ace of hearts and queen of hearts are the most commonly named cards.[9]

The visual saliency force is most closely linked to one of the principles that Derren Brown claimed to be using. Here the magician asks the spectator to mentally choose a card as the magician rapidly flips through them. Each card is only visible for a split second, but one of them, the force card, remains visible for a bit longer. Olson and colleagues have demonstrated that this principle can effectively influence people's choice nearly every single time (98 percent), yet very few notice that their choice has been influenced. Others have reported similar findings, and these results suggest that when people are required to choose from a select group of items, increasing the visual saliency can unconsciously influence their choice.[10]

There are countless other forcing techniques that rely on exploiting people's interpretation of an event. This is the principle underlying the magician's choice force, where you genuinely make a free choice, but the magician frames the selection process in a way that will always result in you choosing the forced item. For example, he might place two cards on the table and ask you to choose one. If you choose the intended card, he will ask you to keep that card. If, on the other hand, you choose the other card, he will ask you to hand it to him, leaving you with the intended card. This principle ensures that you always end up with the force card, and it is often applied using more than two cards. The beauty of this technique is that you do have a genuinely free choice, and as long as you never witness the alternative outcomes, it is very difficult to discover that the outcome of your selection was rigged. Indeed, Hiroki Ozono presented an experiment at the 2017 Science of Magic Association Conference in which participants watched a short video clip in which the magician's choice was used to force one of four cards, after which they were required to work out the method behind the trick.[11] His results showed that only 12 percent of the participants managed to work out the correct solution to this force, and the majority suspected the involvement of a confederate, sleight of hand, or some other psychological principle. The magician's force is an extremely effective and surreptitious method to manipulate your choice.

It is clear from this short discussion that magicians have powerful techniques to influence your choice, and we are starting to learn more about why they work. I use a range of forcing techniques in my performances, and I have experienced the ease with which you can influence someone's choice. However, during a magic show, it is very difficult to ask the audience about how free they felt the selection was. Scientific studies on forcing have revealed that people experience these forced choices as genuinely free. For example, people feel no difference between choices that were genuinely free and those that were forced.[12] However, there is a big difference between forcing a playing card and getting someone to choose a giraffe in Hamleys. Before we fully abandon our sense of free will, let us now turn to mind control techniques in more complex real-world contexts.

MIND CONTROL THROUGH DRUGS AND SUBLIMINAL PERSUASION

There is a long and dark history of scientists and government agencies searching for subversive ways to influence people's behavior. For example, during the Cold War, the Central Intelligence Agency (CIA) set up a top secret research program that aimed to develop mind-controlling drugs that could be used against Soviet enemies.[13] At the time, there was a strong belief that the Soviets had come up with effective brainwashing techniques, capable of altering peoples' thoughts. In fear of losing out on these new psychological warfare tools, the CIA set up Project MKUltra in 1953. This top secret research project, which included researchers from nearly eighty universities, spent the next decade exploring and testing cutting-edge mind control techniques, which ranged from hypnosis to mind-altering psychedelic substances. This research often involved administering drugs, such as LSD, without the subject's consent, after which information on their behavior was secretly gathered. Even at the time, this research was considered highly unethical, so most of the documentation was destroyed, and it is therefore difficult to assess the details of the program. Hallucinogenic drugs and even alcohol can significantly alter our minds, but it is rather doubtful as to whether they can be implemented to control more specific behaviors.[14] Psychologists and government agencies have spent much time and effort researching alternative forms of mind control, and many of them have tried to exploit the idea of tapping into our unconscious mind.

The idea of controlling your mind using unconscious primes dates back to the early days of experimental psychology, and it has always been extremely controversial. The key principle underlying this unconscious persuasion technique, which is also known as *subliminal perception*, involves presenting you with a prime (some visual or auditory stimulus) that you cannot perceive consciously but which will still influence your choice in some way. One of the earliest subliminal perception experiments took place in 1898, a time that preceded our modern projection techniques. The psychologist Boris Sidis studied subliminal perception by showing cards that contained alphanumeric characters to participants who were seated at such a distance that they could barely see the cards.[15] Although most of the participants claimed that they could not consciously perceive the characters, they were able to accurately guess their identity.

In other experiments, Sidis asked participants to choose between two different characters immediately after the subliminal presentation. Although his volunteers claimed that they did not know the identity of the character they had seen, when forced to choose between two alternatives, they performed significantly better than would be expected by chance. Nearly a decade later, Marie Stroh and colleagues demonstrated that you can whisper a name into someone's ear, and although the person is not able to consciously recognize what is being said, they are able to guess the word correctly.[16]

Several other early studies seemed to demonstrate the possibility that our thoughts can be influenced by things we cannot consciously perceive.[17] These findings came at a time when scientists and the public were fascinated with the idea that many of our thoughts and behaviors are influenced and controlled by forces outside our conscious control. For example, Sigmund Freud's highly influential psychodynamic theory relies heavily on unconscious processes, and thus the idea of unconscious mind control found a fruitful seedbed.

The concept of subliminal perception, however, truly captured the public's imagination in the late 1950s, after the publication of a newspaper article in the little-known journal *Advertising Age*. The journal reported a story about a market researcher by the name of James M. Vicary, who discovered a new secret weapon for advertisers: the invisible commercial. His new form of persuasion was based on subliminal perception. Vicary claimed that he was able to flash a commercial for an incredibly brief amount of time—as short as three one-thousandths of a second—and while these messages could not be perceived consciously, they effectively influenced people's consumer behavior.[18] His claim appeared to be based on real scientific data, and he reported that an invisible commercial urging people to drink Coca-Cola and eat popcorn had increased popcorn sales by 57.5 percent and Coke sales by 18.1 percent.

Early in 1958, *Life* magazine picked up the story and ran an article on Vicary's "hidden" selling techniques, which reached a wide readership. *Life* treated Vicary's claims as facts and elaborated on the idea of how this subliminal persuasion technique could be used to change people's behavior more generally, ranging from antilitter campaigns to promoting political candidates.[19] News commentators quickly jumped on the broader implications that this new persuasion technique offered and drew parallels with the totalitarian propaganda in George Orwell's *1984*.[20] These newspaper articles were all published under the backdrop of the Soviet Union launching Sputnik 1, which represented the beginning of the space race and offered the possibility of global remote surveillance and mind control weapons. The concept of subliminal mind control also caught the government's attention, and attempts were made to ban this form of advertising. In the United Kingdom and Australia, subliminal advertising is still banned today.

Although Vicary and his team tried to file for a patent on subliminal advertising, they never actually tested the system, and the results they published in the paper were fabricated. The story that created all of this paranoia was a hoax. Meanwhile, scientists started to question some of the earlier subliminal perception claims by pointing out methodological flaws in their experimental design.[21]

One of the central controversies back then, as is still the case today, surrounded the extent to which the primes were truly invisible. Did the primes really influence people's behavior without them being aware? Charles Eriksen argued that while participants may claim that they did not see the stimulus, they may have glimpsed it and are simply not confident enough to report it when asked.[22] Vicary claimed that his stimuli were presented for less than three milliseconds, and it is unlikely that anyone could consciously perceive a word that has been presented for such a short amount of time. Incidentally, back in the 1950s, the technology was not available for such a rapid presentation. The big problem in much of the subliminal perception research is that if the stimulus is presented for too long, people become aware of it, and it is no longer considered to be subliminal. However, if it is presented too rapidly, it fails to have any influence on subsequent behavior. There is only a very narrow window in which the effects are observed.

Even though there is very little, if any, evidence to suggest that subliminal messages can influence complex behavior, American consumers spend more than $50 million annually on self-help tapes that contain subliminal messages intended to help enhance self-esteem, increase memory, and even lose weight.[23]

Even though Vicary made up his results and the self-help industry unscrupulously exploits people's misconceptions about subliminal perception, this does not mean that subliminal perception does not work.

There are countless more recent studies that have illustrated how subliminal primes can influence our behavior. Neuroimaging techniques, such as functional magnetic resonance imaging (fMRI) and electroencephalography, are providing new insights into how these unconscious primes can influence our brain while being unnoticed.[24] The main debate now concerns the types of behavior that these unconscious stimuli can influence, and it is clear that they are relatively basic. For example, while it is possible to prime someone with an emotional face, it is more doubtful as to whether these types of primes can be used to control more complex behavior.[25]

There are, however, several other techniques that have been used to influence behavior unconsciously. In the previous chapter, we learned about inattentional blindness and the fact that people are blind to things that they fail to attend to. In inattentional blindness, it is possible for people to fail to notice a word, but this word can still influence subsequent behavior.[26] For example, Arien Mack and Irvin Rock demonstrate that even though participants often fail to notice a word presented while their attention is being distracted, this "invisible" word effectively influences their own choice of words.

The idea of controlling your mind through unconscious processes gains further support from studying patients who suffer from a rather common neurological disorder known as hemispatial neglect. This is a neurological attentional disorder that results from damage to one of the two cerebral hemispheres, often as the result of a stroke. Patients suffering from left hemispatial neglect fail to consciously perceive anything that is presented to their left. For example, these patients may only read the left half of a page and simply ignore all of the words that are presented on the neglected side. Neglect can also affect a wide range of nonvisual behaviors, such as dressing and self-care. For example, some patients with visual neglect may only shave one side of their face and only eat food that is located in the nonneglected side. In these patients, there is nothing wrong with their eyes or their perceptual system; they simply cannot attend to that part of the world, making them "attentionally blind." However, even though they cannot consciously perceive things in their neglected field, this unconscious information still influences their behavior. For example, Anna Berti and Giacomo Rizzolatti set up an experiment in which patients with hemispatial neglect were required to respond as quickly as possible to a target picture that was either an animal or a fruit.[27] The patients were also presented with a picture of a different fruit or animal in the neglected field, and even though these images were not consciously perceived, the patients responded much more quickly when the two images were from the same category. This "invisible" stimulus managed to prime their response. These and many other neuropsychological studies add weight to the idea that stimuli we cannot consciously perceive can influence our thoughts and behavior.

All of the unconscious mind control techniques we have looked at so far rely on finely tuned stimulus presentation techniques or brain damage. In 1972, Wilson Bryan Key published an influential and controversial bestseller entitled *Subliminal Seduction*, which suggested there might be easier ways of controlling your mind.[28] Key claimed that advertisers embed sexual images of body parts such breasts and genitals, which, while not immediately obvious to the observer, are picked up by our unconscious mind. Key argued that these hidden symbolic messages stimulate our unconscious mind, which in turn motivates us to purchase the advertised products and brands. The book contained no systematic empirical research but included lots of rather amusing examples. He claimed, for example, that the word "SEX" appeared in the ice cubes of a Gin advertisement, which he saw as clear evidence of this type of trickery.

Key's book was immensely influential and fueled people's paranoia about being subversively brainwashed. Unlike unconscious priming, this form of subliminal persuasion presents fully visible images, but the hidden messages are not consciously perceived. I think that this is the principle most closely related to Derren Brown's claimed method for manipulating Alice to buy the giraffe. Let us therefore look at the evidence supporting this claim in more detail.

Many social psychological theories are based on the idea that people can be primed to behave in certain ways, and in many situations, this priming is incidental and occurs automatically.[29] For example, if you get people to read a set of words that are all related to the concept of kindness, people view others as being kinder.[30] There is little doubt that you can prime someone using a visible prime, but this is not what Derren Brown claims to be doing. Is it possible to prime someone without them noticing the prime? In the priming study described above, all the volunteers were consciously aware of the words that they were reading, and as such, we cannot consider this to be subliminal persuasion. However, more recently, John Bargh and colleagues have published a highly influential and widely discussed study that seems to suggest that the unconscious persuasion of more complex behavior is possible.

Bargh and colleagues designed a clever experiment to test whether words that were associated with being old could influence people's behavior even when they were unaware of the prime.[31] Their participants were told that they were taking part in a language study that involved putting a scrambled sentence into the correct order. For example, the words "piece," "gray," "of," "fabric," and "a" would be reordered to form "a piece of gray fabric." The clever part of this experiment was that, unbeknown to the participants, one of the words always related to being old. They were presented with thirty different words of this type, such as "gray," "conservative," "bitter," and "wise." Much like the subjects in Brown's Hamleys demonstration and Key's subliminal persuasion, the participants in this study were completely unaware of the relationship between the words. Indeed, when later questioned, only a tiny minority reported that they had noticed the concept.

After completing the language task, participants thought that the experiment was over. However, the real experiment was just about to start. The experimenter then measured participants' walking speed as they left the room. Rather remarkably, participants who had been exposed to the old-age words walked more slowly than participants who had not been exposed to these words. Importantly, they appeared to be doing so even though they were completely unaware of the link between the exposure to the words and the behavior. These results seem to suggest that Derren Brown's unconscious persuasion may indeed be plausible. The findings caused quite a stir in the scientific community because they contradicted many of the conclusions based on a large body of research from the unconscious priming literature.

The long history of subliminal perception has taught us to be cautious about extraordinary claims, and indeed, John Bargh's findings were too good to be true. More recently, Stéphane Doyen and colleagues tried to replicate the findings using more controlled measures.[32] Rather than relying on subjective timings, the experimenters used automatic triggers to measure participants' walking speed, and under these more controlled conditions, the priming effect disappeared. These recent findings cast doubt on the possibility of unconsciously priming complex behavior.

There has been a long and controversial debate about unconscious persuasion. While it is clear that some stimuli can influence our cognitive processes, it is important to note that these effects are quite small and often depend on having precisely the right experimental conditions for them to emerge. While research on subliminal perception has important implications for many of our current models of cognition and illustrates that cognition can be influenced without us noticing, these effects are generally too small to be of much practical use in advertising or any other domain.[33] They are simply not sufficiently robust to be applied in any magic performance.

IS FREE WILL AN ILLUSION?

Most forcing techniques do not rely on priming, yet people still feel as if their manipulated choice has been free. At the center of most forcing techniques lies a more fundamental manipulation of free will, and some scientists have started to argue that our sense of free will may simply be an illusion.[34]

Back in 1853, Michael Faraday, who is best known for his pioneering work on physics and electricity, carried out groundbreaking research on people's conscious sense of control by studying spiritualist séances and in particular the phenomenon of table turning.[35] As we saw in chapter 3, spiritualists held séances in which objects would mysteriously start to move. To do so, a group of people would gather around a table and place their hands flat on the table in front of them. The medium would then call upon the spirits and ask for a sign, and in most instances, the table would start to move mysteriously.

Mediums have often been accused of fraud; indeed, many employed tricks such as hidden devices to move objects. However, table turning seemed to be different. Tables appeared to move even in situations where the medium could not have interfered with the table. During this time, many new physical forces were being discovered, and Faraday was curious about the true cause of this mysterious effect. If the spirits could truly influence the physical world, this would dramatically change our understanding of physics.

Faraday came up with an ingenious experiment to investigate the table turning effect. He glued pieces of card to the tabletop with a soft, flexible cement that would give a bit if the sitters' hands moved. He then attached a device to both the table and the card that allowed him to measure the timing of the movements. If the card moved before the table, it would indicate that the movements were initiated by the sitters rather than the spirits. The results were clear: the card always moved first, and thus there was no need to rewrite the laws of physics.

When questioned, the sitters were convinced that they were not moving the table, and most participants were convinced that their hands remained stationary throughout. In a follow up experiment, Faraday used a pressure gauge to illustrate the amount of pressure his sitters were exerting on the table. Once they received feedback about their actual movements, the table stopped moving. Faraday concluded that the sitters were not cheating but that they were using "unconscious muscular actions" to influence the table's movements. Table turning is a very powerful illusion, and it works just as well now as it did in the past. My friend Arthur Roscha often performs this illusion in his magic shows.

Faraday's table turning study was pioneering because it illustrated that people can initiate motor movements without noticing that they are doing so, and there are countless other examples that illustrate these unconscious actions. For example, in the Ouija board, letters are placed around the edge of a board, and sitters are asked to put their fingers on the bottom of an upturned glass. When they start to ask questions, the glass suddenly starts to move without anyone consciously controlling it. While some people argue that the glass is moved through spirit forces, it is in fact moving through physical means, but the movements are initiated by unconscious processes. If you apply pressure to an object, it is difficult to hold your fingers completely still. As our arm gets tired, it becomes increasingly difficult to keep track of individual movements. If someone initiates a tiny movement, others will compensate and adjust their finger positions, which results in bigger movements. In most situations, our bodies continuously adjust position, which is essential to standing on two feet or picking up a drink without spilling it. These small, finely tuned ideomotor movements are part of our everyday lives, yet we rarely notice them. In the case of the Ouija board, this force is amplified to produce astonishing movements.

These ideomotor movements illustrate an intriguing dissociation between our conscious will and our motor actions, and they explain a wide range of ancient mystical phenomena, such as automatic writing, dowsing, and pendulums. More recently, however, Hélène Gauchou and colleagues have argued that Ouija boards can potentially be used to tap into unconscious processes that are inaccessible through conscious reflection.[36] Using a specially designed Ouija board and a memory test, the researchers demonstrated that the Quija board could elicit significantly more accurate responses than a volitional report. As with hypnosis, you are initiating these ideomotor movements yourself, yet you simply do not realize that you are doing so, which is why you experience the movement as if it were controlled by an alien force.

There is much neurological evidence to support the idea that our sense of free will may be an illusion. For example, Wilder Penfield stimulated parts of his patients' motor cortex while they were awake.[37] Since the brain itself does not contain any pain receptors, patients could feel this stimulation. Stimulating specific parts of the motor cortex caused the patients to initiate complex movements that resembled volitional movement patterns. Seeing your hand move without you consciously initiating the movement must be a rather odd experience, and these patients often reported that they did not do the action. Instead, it felt like Penfield had "pulled it out of them." According to Daniel Wegner, these clinical observations illustrate that people's experience of free will may be an addition to the voluntary action rather than the cause of it.[38]

Most of the evidence we have discussed so far is based on neurology patients, and this often makes it rather difficult to generalize to the normal population. However, in 1985, Benjamin Libet published a paper that stirred up debate regarding our sense of free will.[39] As I am writing these words, I feel like I am consciously deciding which finger to press down on the keyboard, and I am pretty convinced that it is my conscious intentions that are causing my fingers to move. Libet set up an experiment to test this intuitive assumption. He asked each volunteer to flex their wrist at random times. Electrodes were attached to their arm and scalp by the experimenters, which allowed them to measure both when the wrist movement occurred and when their brain initiated the motor command. The really hard part of this experiment was finding an accurate measure of when participants felt that the action was being initiated, and to do so, they developed a clever measure. They placed a special clock in front of the participant that had a dot moving around a clock face. All that participants were required to do was to note the dot's location when they decided to move their wrist. Using these three different measures, Libet was able to get independent readings of the time when participants intended to move, the time when the brain initiated the move, and the time when the actual movement occurred.

Libet's results were astonishing: they consistently demonstrated that although the decision to act came about 200 milliseconds before the actual action occurred, the brain prepared the action about 350 milliseconds before the intention to act occurred. What this means is that our brain starts processing the action nearly a third of a second before we intend to act. You might think that a third of a second does not sound like much, but in terms of neuronal processing, this is a significant delay. Even more remarkably, John-Dylan Haynes and colleagues used fMRI to show that the outcome of a decision can be encoded in brain activity of the prefrontal cortex and the parietal cortex up to 10 seconds before people become aware of their decision.[40]

It is hard to envisage that your brain makes decisions before you consciously decide on them but so is the idea that you perceive the future or that your conscious experience is riddled with holes. As we have seen throughout this book, many of our intuitions about our mental capacities are illusory. Likewise, some are arguing that our sense of conscious will is simply an illusion.

Jay Olson and colleagues have recently developed an experimental setup that combines ancient magic techniques with science fiction to further illustrate that our sense of will is rather more malleable than we think.[41] Olson invited volunteers at McGill University to participate in a study that ostensibly tested a cutting-edge mind-reading and insertion technology. Although fMRI can be used to measure specific brain activations, we are far from being able to use these neuroimaging technologies to measure specific thoughts. This was not so in Olson's lab though. He met his volunteers at the cognitive neuroscience laboratory at McGill University and asked them whether they had heard of the Neural Activation Mapping project, which he described as a well-known technology capable of reading and manipulating thoughts. He then proceeded to explain that they were developing new tools to map neural activation patterns onto specific thoughts and were ultimately capable of using this technology to insert thoughts deep into your mind. McGill University is equipped with state-of-the-art neuroimaging technologies but not in Olson's lab. Although the scanner and all of the attached equipment looked impressive, it was all an illusion. The scanner was made of plywood, and the computers for interpreting the neural activations were not even connected to the boxes. All of this was simply intended to convince participants that the researchers were capable of manipulating people's brains.

The experiment started with a calibration procedure in which each participant was required to freely choose a number between one and one hundred and to think of this number while they were lying inside the plywood scanner. Olson then walked back to the data analysis computers and picked up a printout that contained the outcome of the analysis. With the piece of paper clipped to a clipboard, Olson returned to the participant and asked her which number she had been thinking of. The volunteer named the number 41, and while not perfectly correct, the number written on the piece of paper was only off by one. Olson then repeated the same procedure another two times, and on both occasions the number on the printout perfectly matched the freely chosen number.

It is of course impossible to read a person's precise thoughts with a scanner—let alone one made of plywood. All of this was simply theater. Olson used a secret writing device to write the chosen number onto the computer readout after the volunteer named the number, thus ensuring a perfect match on each occasion. Being off by one number was intentional and relied on the "too perfect" theory to make the illusion more compelling. After these compelling illustrations, most of the participants were convinced that the scanner was capable of reading their mind. The most surprising results, however, came when they used the same machine to influence people's thoughts.

Now that the volunteers were convinced that the scanner could read their minds, Olson wanted to see whether the same machine could be used to convince them that it was capable of inserting thoughts. As before, the volunteers were asked to think of a number, but this time, they were told that the scanner was influencing their brain activity in a way that would direct them toward choosing a particular number. After a few moments in the fake scanner, the volunteers were asked to reveal the number that had come to their mind, after which Olson revealed that this was the number that the scanner had sent. To prove the point, he revealed the piece of paper with the matching number.

Again, this was simply an illusion; Olson was surreptitiously writing the number on the printout after it was named. After each of the trials, participants were asked how much control they felt they had exerted over their decisions. Rather surprisingly, participants reported feeling significantly less control over the number that had popped into their mind when they were told the machine was influencing their thoughts. Some participants reported hearing an internal voice that directed them toward specific numbers. The volunteers had a completely free choice, and these findings beautifully illustrate that our feeling of voluntary control is highly malleable and susceptible to top-down influences.

As you reflect on your daily thoughts and activities, it is difficult to abandon the sense of free will that you typically experience. However, there is ample scientific evidence to suggest that this compelling sense of free will may in fact be an illusion. Daniel Wegner has suggested that your experience of conscious will might simply be a marvelous mind trick and therefore an illusory experience analogous to the other illusions we have discussed so far.[42] Wegner has argued that our conscious will may be no more than a rough and ready guide to such a causation, and as such, your sense of free will can be manipulated and misled in many ways. According to Wegner, we experience conscious will when we infer that our thought has caused an action, regardless of whether or not this inference is correct. What this means is that as long as three specific conditions are fulfilled, you will experience conscious control of an action, even if the action has been caused by outside forces: (1) the thought of the action must appear in your consciousness immediately before the action; (2) it must be consistent with the action; and (3) it cannot be accompanied by any conspicuous alternative cause of the action.

Let me explain this theory using an example: Imagine that you are about to flip a light switch. What makes you feel that you consciously intend to initiate this action? According to Wegner's theory, if you think of switching on a light immediately before doing so and if you do not notice anything else that could be causing the light to turn on, you will automatically conclude that you turned on the light after seeing the room illuminate. However, if you suddenly find yourself flipping the switch without having thought about turning on the light beforehand, the lack of consistency between your thought and the action undermines your feeling of free will, and you will not necessarily experience the action as having been initiated by your conscious self. Similarly, if you see another person's hand on a different switch, you might also be inclined to experience less conscious will over having turned on the light. Crucial to this theory, you will experience a sense of agency over your actions regardless of whether or not your thoughts have actually caused the action. Your sense of agency is highly malleable.

In most situations, there is a pretty close correlation between your conscious thoughts and your actions, and thus this correlational approach to free will provides us with a fairly reliable estimate of who's in charge. However, as with our perceptual experiences, there can be huge discrepancies between perception and reality. We are more familiar with perceptual illusions because they can be easily demonstrated in print, but our conscious experience of agency is just as malleable. Let me give you an example that I frequently use to entertain my kids: The next time you are waiting at the elevator and you hear the "bing" sound informing you that the elevator has arrived, use a magical gesture to open the doors. If you get the timing right, the illusion is quite compelling, and as we have learned earlier in the book, our mind is willing to perceive illusory causal connections. This elevator trick often bemuses other observers because the effect is surprisingly enthralling. However, sometimes I even perform the illusion just for myself because, at times, it allows me to genuinely experience a magical force. As I am thinking of opening the doors, I forget about the true cause of the event, and my mind is tricked into believing that it's my mental superpowers that are exerting their influence on the physical world.

Wegner's theory of agency can explain many of the illusions that we have discussed so far, but he reports one experiment that illustrates why people experience the classic force as a genuinely free choice.[43] In this experiment, people were presented with pictures of two objects on a computer screen, and they were asked to use the mouse cursor to select one of them. The volunteers shared the mouse with a confederate, who gently forced the mouse movements without the participants' knowledge. Performed like this, most participants felt that their action had been influenced and thus felt that they had less conscious control over their selection. However, when the experimenters primed the volunteers with a thought that was relevant to the action, the forced movement was interpreted as being initiated by their own will. To do so, the experimenter played a tape recording of the word "swan" about two seconds before they were forced to click on the picture of the swan, and this simple priming resulted in people experiencing the selection as being their own. It is rather remarkable that simply triggering a thought in someone's mind can change the extent to which they experience a forced decision to be their own.

I believe that the same principle is involved in many of the forcing techniques used by magicians. For example, in the classic force, the magician spreads the cards, and immediately before you reach for a card, he inserts the thought that you are choosing a card, when in reality he simply pushes the force card between your fingers. I regularly use this force, and I've been amazed as to why people experience such an obvious forced choice as being genuinely free. Until I started to look into the psychological mechanisms that underlie this force, I underestimated the extent to which people genuinely experience the choice as being free. Indeed, research by Diego Shalom and colleagues has shown that people experience a choice that has been forced using a classic force as equally free compared to a genuinely free choice.[44] However, in light of what we have learned so far about our illusory sense of free will, these results can be explained by Wegner's theory. As long as the thought of freely choosing a card coincides with the card selection, people will genuinely experience the choice as being free, regardless of whether it was forced or not.

So far, we have predominantly focused on thoughts about simple actions, but at times, you have thoughts that do not necessarily involve a physical action. For example, you may prefer blondes to brunettes or strawberry jam to plum. Similarly, you are probably confident in why you will choose to vote for a particular party in the next election. However, although we often claim to know why we make a particular decision, our conscious introspection into the true source of our decisions is often more limited than we think.

Back in 1977, Richard Nisbett and Timothy Wilson published a paper that seriously challenged the extent to which we can consciously introspect about the true nature of our decision-making process.[45] Nisbett and Wilson conducted an experiment to investigate how people reason about the choices they make. To do so, the researchers disguised their experiment as a market research project. They set up a supermarket stall and asked shoppers to rate the quality of stockings. Once they had made their choice, the volunteers had to explain why they had chosen that particular product. As in most psychological experiments, there was a twist: although the shoppers thought they were examining different products, the stockings were in fact all identical.

Rather surprisingly, most of the volunteers came up with elaborate and convincing explanations to justify their choices, but since the stockings were all identical, these explanations must have been fabricated. Moreover, most of the volunteers were completely oblivious to the true cause of their choice: physical positioning. There was a pronounced left-to-right effect in that items on the right were chosen much more frequently than those on the left, but none of the participants claimed to base their decision on spatial location. As you would expect, participants were rather shocked to discover that all of the stockings were identical and even more so that the positioning had influenced their choices. This experiment offered early evidence to suggest that we often fabricate reasons about why we came to a particular decision and that we have poor insight into the true source of our decisions.

The stocking experiment is surprising, but Petter Johansson and Lars Hall used the help of magicians to explore our lack of reliable introspection, which is truly unsettling.[46] The researchers presented their volunteers with pairs of photographs of people and asked them to indicate which of the individuals they found more attractive. Once the participant had chosen their preferred face, the researchers used clever sleight of hand to switch the preferred picture with the face that had previously been rejected. In most of the cases (about 70 percent), people failed to notice that the face had changed. This in itself may seem rather surprising, but based on what we have learned about change blindness, we now know that people are often oblivious to changes in their environment. After the switch, participants were given the face that they had previously rejected, and they were asked to explain their choice.

This is where things turned truly strange, because most of the participants fabricated explanations for their choice. For example, some claimed that they chose the picture because they preferred blondes to brunettes, but we know that this must have been a confabulation because the picture that the person chose in the first place had been a brunette. Rather strikingly, there were no detectable differences in the types of reasoning that participants gave for their real and their false choices, which suggests they were oblivious to their confabulation.

Johansson and Hall have applied the principle of choice blindness to a variety of settings, and the results are surprising and sometimes rather unsettling. For example, in a different experiment, shoppers in a supermarket were asked to taste two types of jam, after which they had to choose the jam they preferred.[47] After revealing their preference, they were encouraged to taste another spoonful of the chosen jam and to verbalize why they had chosen that particular one. As in the previous experiment, the researchers used a magic technique to swap the chosen jam for the one that had previously been rejected. Again, only a small handful of people noticed the switch, and the rest were happy to taste the previously rejected jam while coming up with elaborate explanations as to why they preferred it.

The most dramatic example of confabulations about our choices comes from studies that have looked at how swing voters will confabulate about their political choice.[48] Intuitively, you would think that most voters hold firm political attitudes that they can accurately reflect upon. Lars Hall and colleagues ran an experiment just before the Swedish general election in which they asked potential voters to complete a survey that tested their opinions on some of the key issues that distinguished between the left-wing and right-wing coalitions. Their experiment used clever sleight of hand and trickery to secretly alter the answers on the survey so that the replies were now placed in the opposite political camp. After this sneaky switch, the voters were asked to look back at the survey and explain their attitudes on the questioned issues. As in the other experiments, most of the participants did not notice that their answers had been changed. However, most shockingly, 92 percent of the participants accepted and endorsed the altered political scores. The researchers explained that the survey could be used to identify the political party that most closely reflected their views and opinions. After they summarized the fabricated scores, nearly half of the participants were willing to accept a left-to-right switch. These results are surprising in that they not only illustrate our lack of insight into why we make particular choices but also just how easily those choices can be manipulated.

Magicians have developed astonishing forcing and persuasion techniques, and you are completely oblivious that your choice was anything but free. As with all of the cognitive processes we have looked at so far, forcing relies on exploiting a counterintuitive cognitive illusion: the idea that we have full conscious control and insight into the nature of our behavior. We feel as if we have full control over our thoughts and actions, but much of the scientific evidence suggests that this sense of free will may in fact be an illusion. It is an illusion that is as compelling as the ones we experience through our conscious perception and memory.

Magicians frequently claim to manipulate your mind through scientifically plausible psychological processes. In modern mentalism, spiritualist pseudoexplanations have been replaced by plausible yet often highly exaggerated psychological processes. In recent years, there has been a rather worrying trend in which magicians lecture about psychological persuasion techniques in TED talks and other nonentertainment contexts, which simply fuels people's misunderstanding of the mind. Advertisers and political propagandists have learned to manipulate our thoughts using a wide range of effective psychological persuasion techniques, but these effects are much subtler and less spectacular than those typically demonstrated by magicians.[49]

Derren Brown's unconscious persuasion is a wonderful performance piece, but in light of what we have learned about the human mind so far, there must be more to the illusion than meets the eye. While many other forcing techniques undermine our sense of free will, Brown's unconscious persuasion is implausible if not impossible. Brown is one of my favorite magicians, and I have no intention of revealing how his trick is done (nor the right to do so). As a scientist, however, I can tell you that unconscious persuasion is a pseudoexplanation, rather than the real explanation of the trick. Rest assured that while magicians may exploit your illusory sense of control, we are more resilient to some of the unconscious mind control magicians claim to be using. However, in the next chapter, we will look at a form of mind control that at least superficially gives people complete control over your mind: hypnosis.

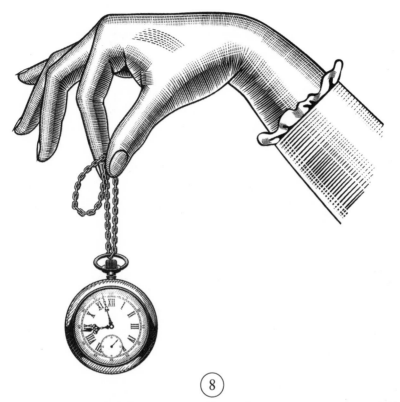

MIND CONTROL THROUGH HYPNOSIS

DERREN BROWN AND HIS TEAM invite a volunteer, Chris, to participate in a documentary about Stephen Fry, one of the United Kingdom's most treasured writers and comedians. Before meeting Fry, Chris is given a small suitcase that contains a real gun, loaded with three bullets. Even though the viewers know that the bullets are blanks, Chris is made to believe that the gun is lethal. However, before he can meet Fry in private, Chris is invited to attend a public lecture, which is being secretly filmed.

Twenty minutes into the lecture, a woman sitting in front of Chris informs him that Stephen Fry is the target, and moments later, Chris reaches for the gun, stands up, and fires three bullets toward Fry. We see blood seeping through Fry's coat as he drops dead onstage. Chris seems unmoved and simply sits back into his chair and falls asleep. The audience is in shock, but to everyone's relief, Derren Brown comes onstage explaining that this was all part of an experiment and that Stephen Fry was unharmed. He continues to explain that Chris is currently under a hypnotic trance and is completely oblivious to everything that has just taken place. Indeed, as the lecture continues, Derren gives a signal for Chris to wake up. Chris awakens and is unaware of anything that has just gone on. In fact, when Chris views the footage of him shooting Stephen Fry, he is genuinely perplexed and fails to recognize his own actions. It is only after Brown activates a posthypnotic instruction to remember all that the memories start flooding back.

Stephen Fry's staged assassination provides a chilling illustration of mind control through hypnosis and, at first sight, seems to undermine the idea of free will. However, before we jump to such a dramatic conclusion, let us look at this form of mind control in more detail and spend some time distinguishing fact from fiction. Fry's staged assassination is one of several compelling illustrations of how hypnotists can take charge of a person's thoughts and actions. For example, in a different event, Brown approaches a passenger on the London Underground and gives him an amnesiac suggestion, and to everyone's amazement, the unsuspecting passenger simply forgets to disembark at his intended stop.[1]

HYPNOSIS ON THE STAGE

Stage hypnotists all over the world have mesmerized audiences by taking control of people's minds. For example, hypnotists might get members of the audience to carry out silly and embarrassing actions, such as behaving like a chicken. Or they might alter people's sensory experiences (taste suggestions), enabling a volunteer to eat a raw onion as if it were an apple. Helen Crawford and colleagues interviewed people after they participated in a stage hypnosis show, and 23 percent of the people questioned felt that the hypnotist had exercised control over them in a way that they were unable to resist.[2] Stage hypnosis certainly perpetuates this idea that the hypnotist has complete control over his subjects, but how much of this is an illusion?

Stage hypnotists provide compelling examples of total mind control, and these demonstrations have molded the public's views on hypnosis. However, stage hypnotists are entertainers, and there may be more to these demonstrations than meets the eye. The first misconception lies in the idea that everyone is hypnotizable. Although some individuals respond very heavily toward hypnotic suggestions, others do not. The volunteers you see on stage have not been selected randomly; the hypnotist screens his audience for highly hypnotizable individuals. For example, hypnotists typically start their show by giving general suggestions, such as that your arm will get heavier and heavier and slowly fall. He will then take note of individuals who respond particularly well and select them for later demonstrations. Thus, the demonstrations you see during a hypnosis show typically relate to only a relatively small proportion of the population, but the hypnotist exaggerates the extent to which hypnotic suggestions can be generalized to the general population.

There is little doubt that stage hypnotists use real hypnosis, but there is an important difference between stage hypnosis and normal hypnosis. The stage hypnotist's primary aim is to entertain his audience, and at the end of the show, the audience needs to believe in his remarkable powers of suggestion. The means by which this is achieved do not necessarily matter, and like magicians, some stage hypnotists use tricks to enhance their suggestions. For example, you may be given a glass of tap water, and after the hypnotic suggestion, the water suddenly tastes disgusting. Although this suggestion may work on its own, a bit of sleight of hand to secretly add some real vinegar will certainly enhance its effectiveness. Other forms of trickery include whispering instructions to the volunteer without the audience noticing. For example, the hypnotist may secretly request you to play along or fake a hypnotic response. I have used these types of instructions in the context of a magic show to provide the illusion of suggestions. I stuck a piece of paper with specific instructions to the back of a box, instructing the volunteer to jump from his chair whenever I touched him. Most people were very happy to comply with my instructions, and the audience witnessed some remarkable suggestions. But of course, this had nothing to do with hypnosis.

Some hypnotists even use magic illusions to deceive the audience into believing that the hypnotist has extraordinary powers. One of the best-known tricks is the human plank, in which a volunteer is given a suggestion that makes his body become completely rigid. The hypnotist then suspends the volunteer's body horizontally between two chairs and dramatizes the effect by standing or sitting on his chest. I often demonstrate this cataleptic illusion in my lectures. It is extremely effective, but it has nothing to do with hypnosis. If you position a person in the correct way they can support much more weight than you would think possible.

Stage hypnosis is not all about trickery, and there are other subtler factors that enhance people's suggestibility. Once the hypnotist has chosen his volunteer, he will ensure that the volunteer endorses the rules of the game, which means that the audience needs to be entertained. This is very similar to volunteers selected to assist in a magic show. Here, there is an implicit contract between the magician and the volunteer that allows the volunteer to be skeptical while essentially doing as he is told. In stage hypnosis, this contract states that the volunteer has the mandate to behave in a manner that under normal circumstances might seem embarrassing and contrary to social conventions. Because the hypnotist usually starts with lots of volunteers on stage, individuals who don't comply with this contract by engaging in the hypnotist's antics are quickly returned to their seats. The context of the stage provides people with a license to act out suggested behaviors without any fear of reprisal. It is therefore likely that some of the demonstrations of loss of control and suggestibility illustrate a person's willingness to comply with the hypnotist's objective to entertain the audience by acting silly.

Our popular understanding of hypnosis has been heavily influenced by hypnosis shows, but these demonstrations may not necessarily reflect the true nature of hypnosis. A skeptic might even suggest that the trickery employed by some stage hypnotists proves that hypnosis is a mere fabrication and, as such, is not a true psychological phenomenon. However, rather than simply throwing the baby out with the bathwater, let us look at the science underlying hypnosis and examine whether hypnotic suggestions genuinely result in a loss of control. Is it possible to control another person's mind through hypnosis and get them to carry out things against their will? Let us start by looking at the origins of this mysterious form of mind control.

EARLY IDEAS ABOUT HYPNOSIS

Over the years, people have come up with a wide range of explanations for hypnosis. Many ancient cultures used therapeutic techniques that utilized incantations to alter people's state of consciousness. For example, there are descriptions of Egyptian healing methods dating back more than three millennia that resemble some of the techniques used in contemporary hypnosis.[3] In the Middle Ages, many of the procedures used to exorcise demons likely involved principles of hypnosis. Even though exorcists claimed that they were using spiritual powers to cure the possessed, it is likely that the effects observed during these procedures resulted from hypnotic suggestions.

The origins of hypnosis are traditionally traced back to Franz Mesmer, a colorful and controversial scientific figure in the eighteenth century.[4] During that time, many new and revolutionary ideas about humans and the universe emerged. In light of this enlightenment, Mesmer built on previous ideas suggesting that there was an invisible animal-magnetic force or fluid that could be harnessed and used to cure physically ill patients. Mesmer specialized in curing patients by manipulating these forces with magnets of all shapes and sizes. For his cures, patients were asked to drink solutions containing iron filings, and he would then pass magnets over their bodies to transmit their energies. Mesmer recognized that as he moved the magnets over limbs, the limbs started convulsing. Mesmer's revolutionary therapies resulted in some rather remarkable improvements and, in some cases, complete cures of diseases that seemed untreatable by more traditional physicians. The news of his successes spread quickly, and Mesmer's magnetic therapies became popular among high-society ladies, who participated in what were considered at the time to be rather risqué magnetism parties.

As the news of Mesmer's success spread, so did the question of how and why people responded to the therapy. More traditional physicians were less convinced by his treatments, and in 1784, a Royal Commission of Inquiry into Animal Magnetism was established to investigate these claims. This commission was headed by Benjamin Franklin and included leading scientists such as the chemist Antoine Lavoisier and Joseph-Ignace Guillotin, a physician and the inventor of the guillotine. The commission carried out well-controlled experiments to investigate

Mesmer's claims. In one of their experiments, Mesmer was asked to magnetize one of Franklin's trees, but instead of bringing the patients to the magnetized tree, the commission brought them to a different one. Another study involved telling the patients that a bucket of water had been magnetized when in fact it had not. The results from these studies showed that people responded according to their beliefs, rather than to whether the objects had been magnetized. Thus, the commission concluded that the effects of magnetism were due to imagination and beliefs. Although the commission's aim was to put a stop to this phenomenon, the practice continued.

Marquis de Puységur, a French aristocrat, was also fascinated by the idea of magnetism and conducted his own experiments in the private laboratories that he had set up in an isolated barn. Puységur mesmerized peasants, who were generally unaware of the convulsions that Mesmer's magnets elicited. Instead of convulsing, one of his subjects fell asleep, and the news of Puységur's sleeping subjects spread quickly. Rather intriguingly, as Puységur's news spread around the country, people stopped convulsing and instead became sleepy when mesmerized.[5]

Ideas of animal magnetism and demon possession have been successfully discredited by modern science. Although the explanations for these behavioral phenomena were incorrect, the behaviors themselves were probably real, and it is likely that people genuinely experienced behaviors that they could not control. The Royal Commission's explanation that people's behavior was due to imagination was intended to dismiss the behavior itself, but the fact that people could experience involuntary behaviors merely through the power of suggestion is immensely interesting and forms the basis of hypnosis. In the last century, clinicians and scientists have tried to explain the mechanisms of this involuntary behavior, and we will now look at some of the theoretical models that were proposed.

HYPNOSIS IN THE LAB

Many psychological theories of hypnosis have been proposed, but the central challenge lies in explaining how hypnotic suggestions can elicit behavior we experience to be outside of our control. One of the main debates has been between *state theories* and *nonstate theories*.

Ernest Hilgard's neodissociation theory is a classic state theory that assumed that hypnotic phenomena are produced through a dissociation of high-level control systems.[6] According to his theory, the hypnotic induction places an "amnesic barrier" between these subsystems and your executive control system, and this dissociation allows for the subsystems to work independently. During hypnosis, people enter a state in which the hypnotic suggestions can act upon the dissociated subsystem. Although people are conscious of the result of the suggestion, the amnesic barrier prevents them from being aware of how the process came about. There are lots of different state theories that differ in detail but essentially propose that hypnosis results in a dissociation between actions and the control of actions.[7]

Nonstate theories question the need to dissociate between different cognitive systems. Nicholas Spanos's social-cognitive theory suggests that hypnotic experiences result from how people interpret the hypnotic suggestion.[8] Accordingly, you experience suggested behaviors as being effortless and outside of your control because you are motivated to interpret hypnotic suggestions as not requiring active planning or effort. The lack of volition therefore comes about when people expect things to be effortless and thus consciously or unconsciously decide to respond to the suggestions. It is important to note that nonstate theories do not imply that people are faking the hypnotic experience. Although people are thought to engage in role enactment or self-presentation, these processes can occur unconsciously, and the unusual experiences are still real.

Stage hypnotists have perpetuated the idea that hypnosis is an effective method to control your mind, but how do we know whether the volunteers are telling the truth? Some dubious practices and tricks have naturally lead to much skepticism, and some argue that the sense of involuntary actions and some of the cognitive suggestions experienced in hypnosis are merely faked. Since the early days of psychology, we have known that people often behave in ways that conform to the demands set out in an experiment. It is possible that people who volunteer to be hypnotized and participants in hypnosis research strive to be good subjects and wish for the experiment to succeed.[9] It is therefore entirely plausible that the hypnotic behaviors we observe on stage, as well as in the research laboratory, result from people striving to behave in an expected manner rather than from the hypnotic suggestion itself.

In most psychological research, people's prior beliefs and expectations about the experiment can influence their behavior, and much care is taken to design experiments that control for these demand characteristics. These effects can be countered through deception that misleads volunteers about the true purpose of the experiment. If participants are unaware of what the experiment is about, they cannot form a hypothesis about it, which removes this source of systematic bias. In other words, the participant must be blind to the experimental hypothesis. The problem with researching hypnosis is that all hypnotic suggestions explicitly inform participants of their intentions, and as such, it is impossible to hide their true objective. For example, imagine an experiment in which you want to investigate whether people can resist a motor suggestion. The motor suggestion only works if the hypnotist explicitly tells you that you cannot move your arm, and the explicit instruction makes it impossible to hide the true intent of the suggestion.

Another way to control demand characteristics is through the use of a control group. In these types of experiments, the control group receives a treatment that appears identical to the experimental treatment. We know from past research that people respond positively to treatment even when the treatment itself is ineffective, a phenomenon known as the *placebo effect*.[10] *Placebo* is Latin for "I will please," and it influences a wide range of clinical symptoms from pain reduction to depression. If you want to discover the true efficacy of a treatment, it is not enough to establish that the treatment works. This is because these results alone do not tell us whether the effect was due to the treatment itself or the placebo effect. The only way to establish the true efficacy of a treatment is to demonstrate that your treatment is more effective than when participants are receiving a placebo treatment. In current medical research, evidence supporting the effectiveness of a therapy is only considered valid if the design has used a randomized control trial that involves an appropriate placebo or control condition. But how can we design an appropriate hypnotic placebo treatment?

Martin Orne came up with an ingenious methodology that allows researchers to control for the demand characteristics that potentially influence people's behavior during hypnosis.[11] His methodology relies on the fact that not all people respond to hypnotic suggestions. Unless you have tried hypnosis, it is impossible to predict whether or not you will respond to the suggestions. We can measure your hypnotizability with a special hypnosis scale, which measures the extent to which you respond to a wide range of hypnotic suggestions.[12] To do so, you receive a ten-minute hypnotic induction, followed by a series of different motor and cognitive hypnotic suggestions, and the experimenter scores the number of suggestions you respond to. Most people respond to some suggestions but not all, a few respond to all suggestions, and very few respond to none of the suggestions. Your hypnotizability does not correlate with any major personality dimensions and is generally very stable over time (twenty-five years). This means that the scores predict how you will respond to suggestions in the future.

The experimenter tests a group of participants who are highly hypnotizable (the *reals*) and compares their behavior with a group of participants who are not hypnotizable but are told to simulate hypnotic behaviors (the *simulators*). These simulators are asked to act as if they were being hypnotized and are told that the experimenter will terminate the study as soon as he discovers that they are simulating, thus encouraging them to act along. Crucially, the experimenter is not informed which participants are simulators and which are reals. The logic behind this experimental design is that if the reals behave in the same way as the simulators, the results must be due to the experimental demands rather than hypnosis. However, if there is a difference between the two groups, we can attribute it to hypnotic suggestibility.

Now that we have an experimental method to study hypnosis, let us take a closer look at the nature of hypnosis. The first step in determining whether hypnotic suggestions are real involves establishing whether hypnotized individuals are telling the truth about their experiences. Several studies have tried to establish whether hypnotized people could deceive the researchers. In one such study, the researchers used a real-simulator design to test whether raters who were trained in detecting behavioral cues could tell whether the simulators or the reals were telling the truth about their suggestions.[13] Although the authors of this study failed to reliably detect any deception in the hypnotized participants, detecting whether or not someone is lying is notoriously difficult, and merely observing non-verbal signals may not necessarily be sufficiently sensitive.

Taru Kinnunen and colleagues used physiological measurements in the form of galvanic skin responses to establish whether hypnotized participants were telling the truth.[14] In the first experiment, the reals and the simulators were asked to either lie or tell the truth about a series of control statements. The galvanic skin responses revealed higher levels of perspiration for both groups when they were lying, thus indicating the measurements were reliably detecting deception. In the second experiment, all of the participants were given hypnotic suggestions and were later asked about the genuineness of these experiences. It is important to remember that the simulators were explicitly instructed to lie about their experiences. The galvanic skin response measurements revealed that 89 percent of the reals met the criteria for telling the truth compared to only 35 percent of the simulators. Galvanic skin responses are not entirely reliable at detecting lies, but these results clearly demonstrate that the simulators who are lying about their experiences were more aroused than the reals, thus implying that the reals were lying less than the simulators.

Skeptics of hypnosis often maintain that the suggestions are not truly experienced but that people are simply complying with the hypnotist's instructions. Social compliance plays an important role in stage hypnosis, but it cannot account for all of the results found in hypnosis research. Irwin Kirsch and colleagues used a real-simulator design to test whether people's hypnotic responses were merely due to social compliance.[15] Both the reals and the simulators were led to believe that they were alone in a room and were given hypnotic suggestions via a tape recording. Unbeknownst to the volunteers, the researchers surreptitiously recorded their behavior using a hidden camera. Afterward, the same hypnotic suggestions were administered in the presence of an experimenter. The results from this study showed that the simulators responded much less to the suggestions when they were not observed, which demonstrates that they were complying with the experimenter. The reals, on the other hand, were unaffected by the presence of the experimenter, and they were just as likely to respond to the suggestions regardless of whether or not they were observed. These results clearly demonstrate that hypnotizable individuals were not simply complying with the experimental demands.

So far, we have established that most hypnotized volunteers are probably telling the truth about their experience and that they are not simply complying with the demands that are put upon them. However, what about a hypnotized assassin, as in *The Manchurian Candidate*? Can we use hypnosis to force someone to carry out such an atrocious act? Derren Brown leads us to believe that this is indeed possible, but there might be other factors at play. Chris knew that he was being filmed, and he may have suspected that the bullets were blanks. It is also possible that the entire episode was fabricated; as we have seen on numerous occasions, it is unwise to blindly trust a magician's explanation. The only way in which we can establish whether hypnosis can turn people into zombie-like assassins is to use a scientifically controlled experiment.

Today's scientific researchers are bound by a code of practice that prevents them from conducting unethical research. Television production companies are not bound by these rules; they can secretly film people's personal lives and create high levels of psychological distress because it is simply used for entertainment. However, back in the early 1970s, the ethical guidelines for psychological research were much laxer. Martin Orne conducted some intriguing experiments investigating the extent to which people would carry out antisocial or harmful acts under hypnosis.[16] Orne was concerned that the behavior observed by hypnotized subjects resulted from their desire to respond as good subjects would and had nothing to do with the hypnotic suggestions. He devised an experiment in which a group of high (reals) and low (simulators) hypnotizable subjects were asked to carry out harmful or immoral acts. In one of the conditions, participants saw an assistant wearing long X-ray gloves place a red-bellied black snake into a compartment. This snake is one of the most venomous snakes in Australia, capable of inflicting mortal wounds unless immediately treated. It is also worth noting that all of these studies were conducted in Australia, where people know about these dangerous animals.

The hypnotist then asked participants to pick up the snake carefully around its middle and place it back in the box. Unbeknownst to the participants, a glass screen had been lowered across the opening, which prevented them from having any real contact with the snake. If anyone would have attempted to reach for the snake, the glass would have stopped their hand. The aim of the experiment was to see whether hypnosis could be used to get people to potentially harm themselves. Rather astonishingly, five out of the six hypnotized reals put their hand inside the box. However, even more surprisingly, so did six out of six of the simulators. Although the hypnotic suggestion clearly resulted in behavior that could have caused damage, this had nothing to do with hypnosis itself and merely reflected people's compliance in following the experimenter's instructions.

What about harming others through hypnosis? In another experiment, the volunteers saw the experimenter pour acid into a beaker, and to demonstrate that the acid was real, he dissolved a coin in the beaker. The experimenter then used misdirection to distract the participant and switched the acid for a harmless solution. The experimenter then asked the subject to throw the "acid" in another person's face. Again, astonishingly, five out of the six reals complied with the instructions. However, so did six out of the six simulators, demonstrating that these harmful acts had nothing to do with hypnosis.

Although the researchers put lots of effort into designing highly realistic test situations, the volunteers did not necessarily believe that their actions would cause any real harm. After the experiment, several of the volunteers said that because the experiment was conducted by responsible researchers, they felt relatively safe and were fairly confident that they would not be harmed. Truly establishing whether hypnosis can be used to manipulate people into doing antisocial acts means leaving the safe environment of the experimental setup. As long as participants know that they are participating in an experiment, they will assume that no real harm can come to them or others.

Back in the early 1970s, a group of psychologists in sunny California investigated whether people would carry out antisocial acts under hypnosis.[17] William Coe and colleagues were aware of some of the limitations of the previous experiments, and they wanted to conduct an experiment in which hypnotic suggestions were used outside the safe confines of the experimental laboratory. The experimenters screened a large number of undergraduates for hypnotizability and then selected a group of highly hypnotizable individuals. As part of this screening process, participants were told that "in the future, when I tell you that now you are hypnotized, you will be hypnotized." This suggestion removed the need for lengthy and elaborate hypnotic inductions for future suggestions. Before participating in the experiment, the students were asked to hand in their schedules and the experimenters used this information to arrange for a "coincidental" meeting with an accomplice whom they had never met before. A few days after the meeting, the accomplice "accidentally" bumped into the volunteer and explained to them individually that he had seen them before in the psychology building and that he would like to talk for a few moments alone.

Once alone, the accomplice employed his short hypnotic induction and explained that his usual partner had not shown up and that he needed the volunteer's assistance to help him sell some heroin. He went on to explain that there was little risk and that he needed the money badly for school and living expenses. He further explained that they would drive to an apartment building together, whereupon the volunteer would take two balloons of heroin (two ounces) to the apartment while he kept watch. Three sharp honks of the car horn meant to get rid of the stuff in the most convenient way, leave by the front exit, and meet him at the end of the block where he would pick the volunteer up. If all was clear, the volunteer was to knock on the apartment door and say she was conducting a survey. The countersign was a response from the occupant that he was from New York but would be glad to help anyway. She should then enter the apartment, give the occupant the heroin, wait while he sampled it, and then collect $100 and return it to the experimenter in the car. After these instructions, the participants were instructed to open their eyes and act naturally while remaining hypnotized. If the subject did not refuse, the accomplice and the volunteer went to the car and performed the act as described above. One half of the participants received these instructions under hypnosis, while the other half received them without.

I cannot imagine that any modern ethics committee would clear an experiment that uses such extreme forms of deception and has such potential to cause psychological harm. It is therefore impossible to replicate these types of studies, but the results are nevertheless surprising. A quarter of the hypnotized participants went through with the crime, and we can be pretty sure that they believed that the staged drug deal was real. Some of the volunteers threatened the experimenters, and the experiment had to be terminated to avoid any serious confrontation. The experimenters explained that

> the first subject, a female, acted as though she were hypnotized and then said she wanted to place her books in a locker in the adjoining building before leaving with the experimenter. On the way, she signalled two male friends who

restrained the experimenter while she ran away. It took the experimenter some time to convince them that it was all an experiment so they would let him go. In the meantime, the subject had disappeared, and they were unable to locate her for several hours. Her father, who happened to be a professor at the university, was told by one of the male friends that something very strange was happening to his daughter. Although the subject was not upset when it was explained that it was an experiment—in fact, she seemed rather amused and pleased with herself—her father was not as easily convinced.[18]

One of the other participants claimed that he intended to double-cross the experimenter and run away with the drugs, as he had been duped by a drug dealer before. The only thing that prevented him from doing so was his fear of reprisal.

These results appear to provide strong evidence that you can use hypnosis to manipulate people into carrying out antisocial acts. However, rather surprisingly, six of the fourteen subjects who were not hypnotized also went through with the crime. The results from the control group clearly illustrate that people's compliance with orders has nothing to do with hypnosis. The idea that we would simply obey a stranger's orders may seem surprising, but back in the 1960s, Stanley Milgram demonstrated that people will obey instructions even if they believe that doing so will cause serious harm.[19] Milgram conducted experiments in which volunteers were given instructions to administer a range of electric shocks to a fellow volunteer. Although the participants were clearly distressed, twenty-six out of forty participants were prepared to administer a deadly 450-volt shock, which clearly illustrates that people will generally obey instructions given by people with authority. In the context of hypnosis, the hypnotist has some level of authority, and this authority alone may be sufficient to make people do things that they would generally consider to be wrong.

Much of the research we have looked at so far illustrates that people are willing to follow instructions, even if this means causing harm to others or themselves, but hypnosis does not add much to people's compliance. It is important to note, however, that this does not mean that hypnosis itself is not a real phenomenon. In recent decades, there have been huge advances in the field of cognitive neuroscience, and neuroimaging allows us to measure brain activation without relying on people's verbal reports. I'm sure we have all told a lie in the past, and despite the fact that we can tell untruths with ease, it is much harder to fake a brain-specific activation. Moreover, just because you are able to simulate hypnotic behavior does not imply that it involves the same cognitive and neural mechanisms. Neuroimaging provides us with new tools to investigate the neural processes underlying hypnosis and is the perfect tool to help us establish whether hypnosis is a real phenomenon.

Positron-emission tomography (PET) is an imaging technique that allows us to identify brain areas that are activated during particular tasks or cognitive processes. Nick Ward and colleagues used this PET imaging technique to investigate whether a paralysis induced by hypnosis would activate different brain areas than when the paralysis was simulated.[20] Highly hypnotizable individuals can be given suggestions that prevent them from bending their arm, and regardless of how much they try, their arm simply will not move. It is fairly easy to simulate a paralysis, but this of course does not prove that the hypnotized subjects have not genuinely lost their sense of control over their actions. The researchers scanned the brains of twelve highly hypnotizable subjects, who either simulated the paralysis or were given suggestions that prevented them from moving their arm. The brain scans clearly revealed different neural activation in the different tasks. During faked simulated paralysis, there were increases in brain activation in the left ventrolateral prefrontal cortex and a number of right posterior cortical structures, which illustrates that the faked paralysis had a different neural basis than the suggested paralysis.

Neuroimaging has also revealed how hypnotic auditory hallucinations activate different neural regions than when words are merely imagined.[21] Highly hypnotizable subjects were presented with the sound of a real word, and they were asked to imagine the word, or they were given a hypnotic suggestion to hallucinate the word. The results from this study showed that an area known as the right anterior cingulate was activated when they heard the auditory word and when they hallucinated hearing it but not when they were asked to imagine the word. Crucially, in this study, the experimenters also scanned the brains of individuals who could not be hypnotized, and the hallucination instructions did not lead to the same brain activation. Similar results have been found for visual suggestions. For example, Stephen Kosslyn and colleagues presented highly hypnotizable volunteers with grayscale patterns and hypnotic suggestions to see color.[22] These suggestions activated color areas in the left and the right hemispheres. Moreover, giving suggestions that implied color images were gray reduced the neural activation in the color areas. These findings clearly support the view that hypnosis is a psychological state with distinct neural correlations and is not just the result of adopting a role.

In recent years, Amir Raz and colleagues have demonstrated that hypnotic suggestions can influence cognitive processes that had previously been thought to be immune to top-down control.[23] As we have seen earlier in this book, many of our behavioral responses are driven by automatic processes. For example, a loud sound or bright light automatically captures your attention, and it is impossible to ignore them. Although some processes are initially automatic, others become automatic through practice and experience. For example, when you learn to drive a car, many of the actions that initially require much conscious thought, such as changing gears, become automatic. Another example of a learned automatic process is reading. Once learned, these tasks become effortless, and it becomes very difficult to inhibit them. As you are looking at the words on this page, it is impossible for you to inhibit reading them. Try looking at the next word but don't read it: magic. It can't be done. If hypnosis can inhibit these attentional processes, it would demonstrate a cognitive ability that goes beyond what is normally possible.

The *Stroop task* allows us to measure this automatic reading process. In this task, you are presented with different groups of words in different colored inks, and you need to name the color ink as quickly and as accurately as possible while ignoring the meaning of the words. Congruent words are words where the color name is the same as the ink color (e.g., "blue" written in blue ink), while incongruent words are words where there is a mismatch (e.g., "blue" written in red ink). To perform this task, you need to identify the color while simultaneously suppressing the irrelevant word's meaning. However, given that word reading is automatic, the word meaning interferes with your task, which slows down the color naming of incongruent words.

Raz and colleagues used posthypnotic suggestions to investigate whether hypnotic suggestions that prevent you from reading, can suppress the Stroop effect. The experimenters told their volunteers:

> Very soon you will be playing the computer game [the Stroop task]. When I clap my hands, meaningless symbols will appear in the middle of the screen. They will feel like characters of a foreign language that you do not know, and

you will not attempt to attribute any meaning to them. This gibberish will be printed in one of four ink colors: red, blue, green or yellow. Although you will only be able to attend to the symbols' ink color, you will look straight at the scrambled signs and crisply see all of them. Your job is to quickly and accurately press the key that corresponds to the ink color shown. You will find that you can play this game easily and effortlessly.[24]

This posthypnotic suggestion abolished the Stroop effect in highly hypnotizable individuals but had no effect on a group of control participants. These results are rather remarkable because the Stroop effect is one of the most reliable effects in cognitive psychology, and it is considered a gold standard for automatic processes. Crucially, the low-suggestibility participants who were given the same posthypnotic instructions were unaffected by the suggestions. These results demonstrate that hypnosis can be used in highly suggestible individuals to inhibit an otherwise automatic process.[25] Current research on hypnosis is uncovering some fascinating insights into the cognitive and neural processes that underpin these mesmerizing experiences. The neuroscience of hypnosis has revealed fundamental differences between the processes involved in experiencing hypnotic suggestions and simulated behavior, which supports the claim that hypnosis is a real phenomenon.

The world of fiction propagates the idea that hypnosis allows you to take full control of a person's mind, and magicians and stage hypnotists have fueled this image. Derren Brown's staged assassination embodies this popular conception and proposes the chilling possibility of manipulating people into doing things against their will. I have tried to provide a critical yet open review of our current understanding of hypnosis and to distinguish between the antics performed by stage hypnotists and the true power of hypnosis.

The scientific research illustrates that hypnosis alone cannot account for why people engage in behavior that appears to be contrary to their own will. Stanley Milgram's disturbing experiments illustrate just how willing we are to follow other people's orders, and there are many harrowing examples throughout history that sadly highlight this unfortunate human tendency. However, the idea that hypnosis can be used to turn innocent individuals into genuine sleeper assassins gains little scientific support, though stringent ethical guidelines prevent us from ever discovering the truth.

Although Manchurian candidates might remain in fiction, hypnosis itself certainly undermines our conscious sense of free will. Central to all hypnotic suggestions is the feeling that your thoughts and behaviors are influenced by outside sources and that you cannot influence these processes. For example, as a hypnotist suggests that your arm is getting heavier, your brain initiates and controls your muscle movements, yet you experience this movement as being independent of your own thoughts. Some stage hypnotists supplement their suggestions with tricks and other forms of social pressure, and some hypnotic behavior can be simulated, but this does not preclude the fact that the hypnotic experience is real. Hypnotic suggestions lead to experiences that feel as if they are outside our own control—it feels like our mind is being controlled—and as such they challenge our sense of free will.

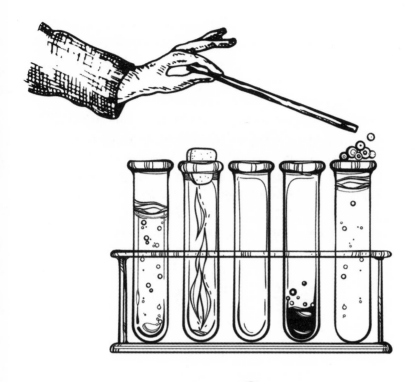

APPLIED MAGIC

IN 1770, WOLFGANG VON KEMPELEN built a machine that had the world in awe. Von Kempelen was an engineer and inventor who, in 1769, had been invited to watch a magic performance at the Austrian royal court. Most of his fellow spectators were spellbound; von Kempelen was not. Inspired by the ease with which his companions were fooled, he promised to return with something that would surpass these feeble tricks. Von Kempelen was true to his word. Less than a year later, he returned with a life-size model of a human head and torso, complete with a black beard, gray eyes, Ottoman robes, and turban. Its left arm held a long Ottoman smoking pipe, and its right lay upon a large cabinet with a chessboard.

Von Kempelen claimed that he had invented a thinking machine, an automaton capable of the most intellectual activity of all: playing chess. He began his demonstration by showing the machine and its parts to the audience while explaining the workings of this mechanical marvel. He opened the doors and drawers of the cabinet, revealing a complex set of gears and cogs that resembled a clockwork. To everyone's amazement, this "Mechanical Turk" could accurately move chess pieces with its arm, each movement being accompanied by a clockwork-type sound. But even more amazing was the fact that the Turk could think: it often won games against the smartest of men, including several grand masters.

News of this revolutionary machine spread like wildfire, and the Turk toured many of Europe's royal courts. Nowadays, we are accustomed to being outsmarted by machines, but at the time, intelligence was exclusively reserved for humans. People had just begun adjusting to the idea that machines could surpass human labor in terms of speed and quality. The Turk now seemed to confront them with another issue that was even more unsettling: could machines actually match or even surpass human intellect?

Today we might be tempted to view the Turk as an early robot that somehow used artificial intelligence. But such was not possible with the technology of the time. Instead, the Turk was an elaborate hoax: Behind its mechanical wheels and cogs was a real person, responsible for the Turk's behavior. The cabinet employed clever optical and psychological tricks that hid the possibility of hiding a person inside. An ingenious mechanism based on magnets and strings allowed the person from within the cabinet to monitor the chess pieces on the board. The invisible helper required candlelight to see, and the Turk's pipe offered perfect misdirection to disguise the candle's smoke. Indeed, one of the reasons so few ever discovered the true secret is that von Kempelen employed clever misdirection and showmanship when he performed the illusion.

Von Kempelen himself was surprised by the impact of his illusion. He often tried to distance himself from it, regularly coming up with excuses as to why it could not perform. But people were hungry for new ideas and inventions, and the Turk offered food for thought. It planted the idea of thinking machines, and many mathematicians and engineers took

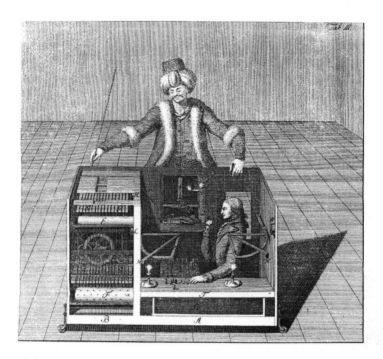

Figure 9.1

Illustration of the Mechanical Turk

inspiration from this magical illusion. Indeed, the Turk inspired Alan Turing, the inventor of modern computers, to try to build genuine thinking machines.[1]

Contemporary magician Marco Tempest follows a similar line by combining magical illusions with technology to give people an impression of what may lie beyond our current technological capabilities. For example, he combines sleight of hand and trickery with state-of-the-art sensing technology to give the illusion of truly interactive augmented reality. Elsewhere, he uses trickery to demonstrate new forms of human-robot interaction.[2] Such demonstrations allow us to experience technologies that are currently impossible but may inspire engineers to make such impossibilities possible.

Magic can therefore affect our lives beyond just performance. We have already seen how spiritualists exploited magic, and we have looked at how anomalous experiences can influence our beliefs. In this chapter, we will explore how magic can be further applied beyond the stage, often in rather surprising areas.

MAGIC AND HUMAN-MACHINE INTERACTION

Have you ever written a personal email or updated your social media status while at the office? People are often reluctant to do so because they do not want their employers or colleagues to notice that they are spending work hours on personal matters. A group of Korean computer scientists came up with an ingenious solution to this problem by designing a desktop interface that uses a decoy text window to draw the attention of passersby away from their personal emails.[3] As an employee writes their personal email, the preprogrammed decoy text appears in a prominent window, giving everyone else the impression that the writer is busy with a serious piece of work. Even colleagues who were warned that suspicious activity might occur rarely noticed this surreptitious move. I myself rarely feel the need for such devices, but this example nicely shows that the link between magic and technology can often be closer than most of us think.

"Any sufficiently advanced technology is indistinguishable from magic," wrote Arthur C. Clark.[4] I vividly remember the first time I interacted with an iPhone's touchscreen, and the feeling was indeed magical. But the connection with magic goes deeper. Bruce Tognazzini points out that much of successful human interface design intentionally draws upon principles of magic, with designers and magicians both creating virtual realities that are based on illusion.[5] For example, we click on virtual files on our desktops; I just deleted an earlier draft of this chapter by dragging its file to a trash can icon. But in reality, no such trash can exists—this is just as much an illusion as David Copperfield's body being cut in half. Indeed, designers in the area of human-computer interaction explicitly call this the *user illusion*.

Such illusions are also pivotal to many of our interactions with computers. For example, in the 1990s, being online was rather special, given that internet use was charged by the minute. (As a cash-strapped student, I used it rather sporadically.) The advance of all-digital telephone systems enabled calls to be connected within milliseconds and data exchanged in brief high-density bursts. Such responsiveness has allowed telecommunication companies to save money by going online only during the milliseconds when actually needed, while maintaining the illusion of constant connection.

These smooth human-computer interactions often rely on various principles developed by magicians. For example, magicians often give the impression that what they are doing is much more difficult than it really is. Harry Houdini was a master at this, often giving the impression that his escapes were incredibly difficult. After Houdini was handcuffed and tied up, spectators usually had to wait for several minutes before he emerged free. The escape itself took only a few seconds, but Houdini spent the time intentionally waiting (often reading a magazine) before sprinkling himself with water and bursting forth, looking properly sweaty and exhausted. This same principle is now applied in many software designs.

For example, back in the mid-1970s, Fairchild produced one of the first home video game machines. It featured a first-rate tic-tac-toe game that had a flaw: regardless of how long a player took to plan a move, the computer responded immediately with the best possible countermove, leaving the user feeling puny and inadequate. This is one reason why Fairchild no longer makes video games. And it is also why subsequent games often include artificial delays, giving you the illusion that the computer is spending valuable resources planning its next move.

Such benevolent deception is used in other technologies as well. For example, have you ever wondered what truly happens when you press the button at a pedestrian crossing? Although some signals respond to your press, others don't. By 2004, the city of New York had deactivated its pedestrian buttons, as most of its traffic lights were computer controlled. Nevertheless, millions of people still press the nonfunctioning "placebo" buttons with no additional benefit beyond the illusion of having control.[6] Similar examples include thermostats that hiss when pressed, a sound that serves no other function than to provide you with an illusion of control.[7]

Even though magic principles may offer the prospect of more fluent (and fun) interaction with machines of various kinds, there is also the danger that they will be exploited for malevolent ends. Back in the 1960s, an enormous piece of hardware—the electronic switching system— managed connections between phone lines. Despite being reliable most of the time, there were failures, and the system knew when such failures had occurred. But rather than disconnecting, the switchboard purposefully connected the caller to a wrong person. This misled callers into believing that they had dialed a wrong number, while keeping intact the illusion of an infallible phone system.[8]

As we become increasingly dependent on technology, we also become increasingly susceptible to technological deception. The 2016 US presidential elections have highlighted the extent to which technology can be used to manipulate public opinion. Bots spread fake news across social-media platforms, creating the illusion of authenticity. Programmers in Silicon Valley conjure up targeted propaganda on our computer screens with far more secretive algorithms than any magician's secret I know. It is therefore imperative that we become aware of how far such deception can go.

MAGIC AND REAL-WORLD DECEPTION

Although magic necessarily involves deception of some kind, deception is not unique to magic; it is part of many of our everyday activities, ranging from competitive sports to business interactions. Deceiving others is central to much of life on our planet; it can be found in most animals and even plants.[9] Given that magicians are masters of deception, it is not surprising that many of their techniques have been applied to other domains. The psychology of deception covers a huge body of research, and it is not my intention to discuss it all here. Instead I would simply like to shine a light on the symbiotic relationship between magic and deception in the real world by focusing on some rather surprising areas of application.

Relatively little is known about the beginnings of magic. But it is fair to assume that magic originated from various forms of deception. Many of the tricks performed today evolved from street games in which misdirection was used to swindle people out of their money. For example, the Cups and Balls is based on the Shell Game, an ancient game still played on many streets today. This involves three objects (such as shells, bottle caps, or plastic cups) laid out on a flat surface. The hustler places a small ball under one of the shells and then quickly shuffles them around the table, after which he asks for bets on where the ball is.

The Shell Game involves multiple layers of misdirection and, when performed well, represents a true masterpiece of deception. First, the operator uses sleight of hand to secretly move the ball from one shell to the other. I have seen this scam performed many times, and even my trained eye often fails to detect the trick. But there is more. The game often involves several individuals working as a team. These teammates continually win, causing the casual observer to believe that winning is easy. After watching a few such games, the naïve punter, excited by the riches being made, becomes tempted to try his own luck. He might try a few times using small bets and wins. But don't be fooled; these winnings are part of the deception, intended to instill a false sense of security. His fellow players soon egg him on to increase the bets. He might even spot the hustler cheating, and while his "fellow" players are betting on the wrong shell, he now feels confident in knowing the ball's true location and so raises the stakes. But this is a double bluff: as soon as larger amounts of money become involved, the tables turn, and the hustler now uses sleight of hand to ensure that the money flows into his own pockets.

If you look very closely at Hieronymus Bosch's fifteenth-century painting of the Cups and Balls (figure 4.1), you can spot a small boy pickpocketing the punter—a vivid illustration of the symbiotic relationship between magicians and hustlers. Please let me be clear: I am not accusing magicians of being dishonest. Rather, I am saying that historically there has been a close relationship between magicians and swindlers such as pickpockets, cardsharps, and even shoplifters.[10] Many of the card sleights and gimmicks used by magicians were originally developed by gamblers stacking the odds in their own favor. I am likewise confident that many techniques developed by magicians have found their way to the poker table. This is true of gimmicks as well. For example, the Topit allows you to toss objects into a secret pocket in your coat, providing a powerful tool to vanish things. This was likely inspired by a device used by shoplifters known as the Poacher's Pocket. I share Derren Brown's view that many teenage magicians have been tempted to use magic principles to sneak objects into their own pockets as well.[11]

Let us now look at deception in more "respected" circles. French illusionist Jean-Eugène Robert-Houdin was one of the most influential magicians of all time. By performing elegant and sophisticated magic, he brought magic to the upper classes of society and, in so doing, distanced it from crooks and swindlers. Offstage, Robert-Houdin claimed to be involved in an even grander venture: political deception. In his memoirs, Robert-Houdin recounts how, in 1856, the French government enlisted him to help quell an uprising in Algeria.[12] The French were losing control of the region, due in part to marabout sorcerers who put on enthralling displays of magic to encourage local leaders to rebel. A French colonel had the idea to fight magic with magic; he engaged Robert-Houdin to put on a show that would rival those of the local sorcerers. With the help of translators seeded throughout the audience, Robert-Houdin created powerful illusions to impress and intimidate the local tribes. For example, he produced cannonballs from a top hat to demonstrate France's limitless armaments. He performed a version of the infamous Bullet Catch, in which he used his teeth to catch a bullet that had been fired at him. And to top it all, he created his most ingenious illusion, the Heavy Box.

In this illusion, Robert-Houdin asserted that he had the power to "deprive even the most powerful man of his strength." Anyone who doubted him should come forward. A muscular Arab typically responded to the challenge and, at Robert-Houdin's request, easily managed to lift a wooden box from the floor. After this demonstration Robert-Houdin proclaimed that "now you are weaker than a woman: try to lift the box." The man then grasped the handles, exerting all his force, but without effect. And on his final unsuccessful attempt, for no apparent reason, he shrieked, fell to his knees, and ran from the stage.

Although Robert-Houdin gave the impression of having supernatural powers, his illusions were carried out via simple tricks. For example, what appeared to be a normal wooden box was far from normal: its bottom contained a steel plate, so that when an electromagnet concealed under the stage was turned on, the box became impossible to move. As to the cry of agony, this was achieved by sending a painful electric current though the metal handles. Robert-Houdin himself claimed that his magical interventions were instrumental in suppressing the local tribes.

Magicians also have much in common with the military, in that both groups aim to distort someone's perceived reality using similar techniques. For example, both groups use camouflage and feinting to hide their true actions and intentions.[13] One person who combined the two areas was Jasper Maskelyne, a successful British magician who helped the Allies defeat Axis forces during the North African Campaign. It is claimed that he led a "magic gang" to create tactical diversions, which included camouflaging the city of Alexandria, Egypt, from aerial bombardment and creating phantom tank battalions that acted as decoys. Maskelyne's heroic tales have been recounted in films and books, but (as has also been said of Robert-Houdin) these stories may be more fiction than fact.[14] Although there is little doubt that Maskelyne played an active role in the North African Campaign, he probably deployed existing camouflage and decoy technologies rather than the exaggerated magic-inspired innovations. The military itself may have deliberately exaggerated his role and used his celebrity status as a propaganda tool. As Jonathan Allen points out, "With a 'Maskelyne' in charge of deception operations, whatever could the German High Command expect next?" The British government may have used a magician as an illusory force.

However, not all the involvement of magicians in war and politics can be relegated to the realm of fiction. Back in the 1950s, the Central Intelligence Agency (CIA) explored a wide range of espionage methods, and it may come as no surprise that they had a keen interest in magic. As part of their effort, the CIA asked one of America's best-known magicians, John Mulholland, to create a manual to help their agents improve their deception skills. These documents have now been declassified and published.[15] They make for interesting—and often amusing—reading. For example, Mulholland describes different deceptive strategies that agents can apply. One of my favorite techniques, designed to lose your tail, involved deploying a spring-loaded puppet on the back seat of your vehicle. This gives the illusion that the agent is still in the car, when in reality he has slipped away on his secret mission. The manual also contains a large section describing different misdirection techniques for covertly administering toxins in the form of pills, liquids, or powders into another person's drink—misdirection that kills.

Magicians have always been entangled in a wide web of secrecy and deception, which often makes it difficult to distinguish fact from fiction. However, with the recent advances in our knowledge of magic, scientists and magicians have started to collaborate more openly, allowing for a more transparent transfer of knowledge. The topic of deception is naturally of great interest to people working in cybertechnology. Knowing the multiple layers of deception used by magicians can help us design computer systems that are more resilient to cyberattack.[16] Along these lines, Wally Smith, Frank Dignum, Liz Sonenberg, Michael Kirley, and I have just been awarded funding by the Australian Research Council for a project to implement magicians' principles of deception in autonomous computer agents.

Politicians are often accused of deceiving the public and using political misdirection to keep awkward issues hidden from public view.[17] Spin doctors and political strategists can deploy misdirection on a massive scale, manipulating the media to manage public perception of their candidate. Lynton Crosby, for example, is one of the most successful political advisers of all time and a true master of political misdirection. One of his signature maneuvers is the "dead cat," which British politician Boris Johnson has described as follows: "There is one thing that is absolutely certain about throwing a dead cat on the dining room table. ... [E]veryone will shout, 'Jeez, mate, there's a dead cat on the table!' In other words, they will be talking about the dead cat—the thing you want them to talk about—and they will not be talking about the issue that has been causing you so much grief."[18]

These forms of political misdirection are extremely effective, and politicians scrupulously exploit the ease with which the public can be misled. For example, on March 4, 2017, Donald Trump famously sent out a tweet accusing Barack Obama of wiretapping him during the election campaign. Despite being unfounded, as with the dead cat, such claims are difficult to ignore. The world media discussed the issue for several days, leaving other topics of his troubled administration in the shadows.

Spin doctors, advertisers, and magicians all manipulate our beliefs by systematically orchestrating our experience. As with many of the other forms of deception we have looked at so far, these professions have many of the same tricks and general techniques. They can potentially learn a lot from one another.

MAGIC AND FILM

The words "magic" and "film" conjure up images of Disney magic, because magic is deeply ingrained in the Disney brand. However, at the beginning of the twentieth century, magicians brought a new type of magic to the screen. The turn of the twentieth century was the golden age of magic, and magic was deeply engrained in popular culture. Before the rise of cinema, live performances were the main staple of popular entertainment. Magicians often incorporated the projection of moving images into their programs as standalone films or backdrops to their performances. In some cases, they seamlessly merged live performance and film projection, bridging the boundary between the physical and virtual realities.[19]

In the early days of cinema, magic was frequently featured in these short films, and people would watch in amazement as a conjurer performed sleight of hand in one continuous, unedited shot. For example, one such film from 1896 shows a magician performing the linking rings and a card trick and then producing eggs from an assistant's mouth.[20] These early films simply recorded magicians performing tricks in the same way they performed them on stage. But magicians and filmmakers soon realized that this medium had the potential to elevate the art of illusion to a new level.

Georges Méliès became fascinated by magic after seeing Jasper Maskelyne perform in London. In 1888, Méliès bought the Théâtre Robert-Houdin, where he regularly performed, but predominantly worked backstage to help design new magic illusions. In 1895, he attended a demonstration of the Lumière brothers' cinematograph. Inspired by this revolutionary device, he built his own projector and designed a film camera. In 1896, Méliès created the first cinematic studio, allowing him to bring his experience in misdirection and stage illusion to the screen. For example, in his film *The Disappearing Lady*, a magician walks out onto the stage with a female assistant and spreads a newspaper on the floor to illustrate that there is no trapdoor on the stage (figure 9.2). He places a chair on top of the paper and sits his assistant in it. The magician then spreads a blanket over his assistant. When he removes it, the lady has vanished. When performed on stage, this trick would have required a trapdoor, and the transformations would have been far less smooth. But Méliès's clever editing tricks allowed him to create a powerful new form of illusion.

Rather than using sleight of hand to crate illusions, these short films employed cinematic techniques such as substitution splices, multiple exposures, and reverse motion. These trick films delighted early cinema audiences around the world, and it was assumed by early filmmakers that

Figure 9.2
Méliès's Vanishing Lady

such tricks were needed to hold the audience's attention. But this assumption did not stand the test of time; by the second decade of the twentieth century, trick films had been replaced by documentaries and films with stories and narratives. Méliès did, however, manage to leave two lasting legacies: a new area of cinematography called special effects and a new film genre called science fiction.

Magicians played an important role in pioneering new cinematographic techniques and helped promote this new form of entertainment. It is therefore ironic that the moving pictures replaced the public's thirst for live entertainment and with it the demand for live magic.[21]

Although most special effects in film nowadays are created through computer graphics, the special effects in live performances, however, are still based on magic. For example, Paul Kieve, one of the United Kingdom's most innovative magicians, has developed stunning illusions for the musical *Ghost*, in which a real person seamlessly disappears into thin air.[22] Magic illusions are also common in amusement parks that provide immersive experiences. For example, Derren Brown's Ghost Train, a ride at one of the United Kingdom's leading amusement parks, uses clever illusions to frighten people nearly to death.

MAGIC AND DESIGN

As the focus of home entertainment shifts from passive viewing to interactive experience, pressure is increasing for game designers to create games that effectively engage an audience. One of their main challenges is to get preprogrammed avatars to engage in genuine social interactions—a problem magicians could help with. Most magic performances appear to involve lots of spontaneous social interactions but, in reality, have a considerable amount of fixed structure. For example, as I perform the Cups and Balls, it appears as if I am having a genuine social interaction with my audience. Yet every move and word has been carefully scripted. The spontaneity and interactive nature of such interaction is another powerful illusion. Many of these principles could also be applied in video games to give the illusion of genuine social interaction.

Many other magic techniques have yet to be explored in this context.[23] For example, forcing could be applied to interactive games, giving players the illusion of choice but, in reality, leading them toward a predetermined goal. For example, one of the classic *Zelda* games used a (rather annoying) fairy named Navi as a way of redirecting the player's attention back to the predefined goal of saving the land from evil. This prevented players from getting sidetracked in a seemingly free and expansive virtual world. Forcing techniques would allow game designers to steer the game less intrusively. Similarly, misdirection could be employed in virtual environments to ensure that new players would see what they need to see, thereby unobtrusively guiding them through the game. Several illusion techniques might also be employed to make visual objects seem more real and to help bridge the gap between the virtual and the real worlds.

For example, in many virtual reality games, players are fully aware of the VR headset, which can create a physical barrier between the physical and the virtual worlds, making such games rather unengaging. Derren Brown's Ghost Train uses a clever trick to bridge this gap: the riders are asked to wear a gas mask (actually a VR headset) to protect them from poisonous gases released into the train. This explanation accounts for the presence of the head set and so changes their framing of it, helping riders immerse themselves more fully in the VR environment.[24] There are countless other parallels between game design and magic. Early filmmakers learned a lot from magicians; the same may also be true for modern game designers.

Magic techniques are used as the basis for design in other areas as well. For example, magicians often use clever combinations of color, shape, and texture to make boxes appear different than their true physical properties.[25] Although I am not aware of magicians directly consulting with architects, similar illusions have long been used to enhance a building's aesthetic appeal. Greek and Roman architects, for example, adjusted the proportions of temple columns so that these would appear straight when viewed at a distance.[26] Similarly, the floor of the Parthenon is purposefully curved upward because a flat floor would appear to sag inward. And although newer engineering techniques and materials give contemporary architects much more flexibility, there still remain physical limits as to what can be built. But by misdirecting the observer's attention away from load-bearing structures and using illusions to disguise the true size of a structure, constructions can appear to defy the laws of physics. If done properly, the resulting buildings can be visually stunning.[27]

MAGIC AND WELL-BEING

People rarely think of magic as a way to enhance well-being, but such applications do exist and, in fact, are rapidly growing in popularity. Until now, this endeavor has involved a rather disparate group of techniques, but Steve Bagienski and I have recently developed a hierarchical framework to help organize many of these approaches (Figure 9.3).[28]

Let us start by exploring some of the positive effects on our well-being caused by a magic performance. People are strongly captivated by magic, and most of us enjoy the feeling of wonder it elicits. Until fairly recently, people considered magic as simply another form of entertainment. However, new developments suggest that our fascination with magic connects to something more intriguing: a deep-rooted adaptive cognitive process that encourages us to keep learning about the world.

Our interest in the impossible starts at an early age. Young infants typically look much longer at events that violate their understanding of the world. This response is so reliable that it is one of the main tools used to study infant cognition. An interest in things that you cannot explain

Figure 9.3

Progressive hierarchy of well-being effects in magic

also has many adaptive advantages, such as encouraging you to explore your environment and better understand your world.[29] Consistent with this view, Eugene Subbotsky has shown that children are attracted to situations that they deem impossible (i.e., magical).[30] This sense of wonder that magic elicits appears to serve an adaptive function, as is the case with most emotions. I believe that we become captivated by an apparently impossible event because it pushes us away from the mundane knowledge that we have already mastered and toward the unknown, thereby helping us to expand our knowledge of the world, which may prove useful later on.

If this view is correct, it might also be possible to harness this expansive interest in order to become more creative. This theory is speculative, but there is evidence that experiencing magic encourages people to explore things from new angles. In one such study, children watched movie clips from a *Harry Potter* film that either did or did not contain strong magical content.[31] Immediately after watching these clips, the children completed several standard creativity tests. Those who had watched the clips with the magical content were significantly more creative than those who had not. Building upon this work, Arthur Roscha, Rianne Stewart, and I are currently investigating whether seeing magic tricks in real life can make adults more creative. The results so far have been mixed.

Because seeing a magic trick triggers a strong need to know how the trick was done, this need can also be harnessed to enhance learning. For example, Peter McOwan and colleagues are using magic to motivate children to study science, technology, engineering, and mathematics.[32] Others have taught computer-programming skills by using a mind-reading trick where students are given a table of various numbers and mentally choose one.[33] After telling the instructor only which columns it is in, the instructor reveals the student's chosen number. The students are then encouraged to discover how the trick works as the trick is repeated. The method (eventually revealed by the instructor) involves a principle used by Hamming Codes, which is precisely the focus of the lesson. The evaluations this course has received are extremely positive, suggesting that magic may be a useful teaching resource.

One of my colleagues, Matt Pritchard, performs magic in schools across the United Kingdom with a specially designed magic show that encourages children to learn about science.[34] He does this by performing relatively simple magic tricks, after which he asks the kids how the tricks were done. Our intrinsic need to work out tricks is a natural way to get children thinking more critically and to encourage scientific thinking in a fun and engaging environment. The creativity boost elicited by the magic itself may also enhance children's problem-solving skills. In addition, magic is simply fun, and Matt's magic shows put children in a better mood, which probably helps them learn more efficiently as well.

As you can probably guess, I frequently use magic tricks in my lectures. In fact, student feedback shows that my students want even more magic (and probably less cognitive psychology). Many of my colleagues tease me about teaching magic rather than psychology, although I sometimes spot small glimmers of envy. In any event, simply performing a few tricks in your lecture won't automatically get your students to learn more. In a recent study where participants either watched a video clip of a magic trick or a circus act before a neuroscience tutorial, researchers speculated that because the experience of magic activates brain areas underpinning working memory, seeing magic before the tutorial could enhance learning.[35] However, this was not the case: seeing magic before the tutorial made them engage less, rather than more. Results also showed that not knowing how the trick was done made students ruminate, thus distracting them from learning neuroscience.

It is important to note that the magic trick in this study had nothing to do with what the students were required to learn. In contrast, I rarely perform magic tricks simply to spice up my lectures; the tricks I perform usually deal with the topic at hand. For example, I might use a misdirection trick when teaching about attention or a trick based on illusion when teaching visual perception. Pritchard likewise uses tricks that have been specially designed to encourage problem solving. Even though magic tricks may not be a magic bullet for learning, I am convinced that when applied correctly, they can help people better engage and learn complex material. However, we need more research to find the best ways in which this can be done.

Watching magic tricks has also been shown to help manage pain and anxiety. For example, Gionatan Labrocca and Edda Oliva Piacentini used magic tricks to distract children before having their blood taken, which significantly reduced the amount of pain the children experienced.[36] In one of my favorite studies, researchers recruited the help of a group of clowns who performed magic tricks to reduce anxiety in children about to undergo surgery.[37] The clowns accompanied the children and their parents into the operating room, performing tricks until the anesthetic was applied. This use of magic did indeed reduce the children's anxiety, but staff were concerned that the clowns interfered with the rigid medical standards of the operating room. On reflection, clowns and medicine may not have been a perfect mix. Dentists have likewise dealt with strong-willed children by performing magic tricks before sitting them down in the dental chair.[38] The tricks indeed made the children more cooperative, and researchers suggest that the magic might have helped these children perceive the dentist as a playful and approachable ally, thus reducing anxiety.

There has also been a long tradition of enhancing physical and psychological well-being by teaching people to perform magic. For example, Breathe Magic is a United Kingdom–based organization that developed a training program to help children with hemiplegia (a neuromuscular condition that prevents people from using one side of their body). These individuals find it very challenging to do simple bimanual tasks that we typically take for granted (e.g., tying a shoelace). Several physiotherapy programs are able to help these children, but most children are poorly motivated to engage in these programs, which involve repeatedly carrying out painful actions. Breathe Magic, however, utilized professional magicians who collaborated with physiotherapists to design special magic tricks that encourage children to use both hands. The results were remarkable. After a two-week summer magic camp, the children significantly improved in many bimanual tasks.[39] An unexpected bonus also appeared: children with hemiplegia often struggle to integrate properly into mainstream education and can become socially withdrawn. But learning magic tricks seemed to give them a new confidence in socializing and to help their self-esteem.

Teaching children how to perform magic tricks might also improve learning for students with learning difficulties. For example, Kevin Spencer developed a Hocus Focus curriculum that motivates such students by having teachers integrate the learning of magic into their lesson plans.[40] Spencer suggests that the tricks themselves create a strong sense of curiosity and improve self-esteem by giving students skills that their nondisabled peers do not have. Teachers report that magic tricks also help students to actively engage with the content and to better develop their attention and problem-solving skills. Although more research is still needed before we can draw firm conclusions, these results are encouraging. In collaboration with Abracademy, Steve Bagienski and I are currently studying whether magic can be used to help children with low self-esteem and poor social skills to overcome these limitations.[41]

Magic has great potential to help a wide range of individuals, and as the science of magic gains momentum, its use to enhance well-being will likely increase. For example, my colleague Darren Way has used street magic to build rapport with youth who typically develop destructive lifestyles that include gangs, drugs, and crime.[42] Darren suggests that people who are drawn to gangs are likely to be attracted to magic, because the lifestyles of both magicians and gang members involve showing off and gaining credibility by using deception. The main difference, however, is that magicians are benign and harmless. Darren also founded the Streets of Growth charity, which has used magic to connect with gang members on the streets of London for nearly twenty years. He has also personally used magic to disarm gang members and resolve conflicts. And most importantly, magic enables Darren to connect with disenfranchised individuals and allows them to engage with his charity and other services, which then helps them transition toward a more promising future.

There are lots of different ways in which magic is being used to enhance well-being. It does not provide a magic bullet on its own, but I strongly believe that it provides an immensely useful—and often neglected—tool to complement other therapeutic and pedagogical approaches.

A SCIENCE OF MAGIC: ATTEMPT #1

In the nineteenth century, magicians took great inspiration from the latest scientific discoveries and often astonished the public with demonstrations similar in many ways to those being given by scientists.[43] Many magicians also had backgrounds in science and worked as engineers, inventors, and skeptics, defending science against the deceits of the mediums and other psychic tricksters.

Likewise, many scientists of that time were intrigued by magic and used their scientific skills to explore apparently supernatural powers. As we have seen earlier in this book, Michael Faraday developed a mechanical device to study table turning, while Richard Hodgson and S. J. Davey conducted experiments to discover why people misremembered the things that occurred during a séance.[44]

Working with mediums was not easy because they rarely explained how their effects were achieved, and their supernatural powers were rather unreliable. Stage magicians offered a more attractive proposition; they could explain the supernatural aspects of their illusions, and they were in full control of their abilities. Many eminent scientists at the time realized that conjuring offered a valuable way to uncover how the mind works, and this was the beginning of the science of magic—a scientific endeavor that saw magicians actively participate in scientific research, renouncing the secrecy of their trade for the scientific enlightenment.

Scientist have always been fascinated by geometrical and optical illusions, and by the end of the nineteenth century, they became interested in studying the illusions created by magicians. These conjuring illusions offered a new way of studying human experience and provided new insights into our consciousness. Many researchers present at the beginning of psychology—such as Max Dessoir, Joseph Jastrow, and Alfred Binet—realized that a scientific understanding of magic could provide important insights into the human mind.

Dessoir was a German philosopher, psychologist, and amateur magician. He studied magic books and performed some tricks himself, in the hope of understanding the mechanisms of the mind. He was convinced that psychologists could learn from magicians. By analyzing some of the basic principles in magic, he developed several interesting theories, many of which have withstood the test of time but have sadly been forgotten by contemporary science.

One such theory stemmed from Dessoir's observation that if a pretend action occurs after a real one, the pretend action is often seen as real. For example, imagine a performer at a table who throws an orange up into the air and catches it on its return with one hand, which drops a little bit under the table. He repeats this action again. Next, he leaves the orange in his lap and only pretends to throw it. Dessoir claimed that nine-tenths of the audience would see the orange rise and disappear in midair (although he did not back this up with empirical data). He used this effect to suggest that mental images can be experienced just as real percepts, even without a corresponding external reality. Such "positive hallucinations"—which had previously been observed only in the mentally unwell—could therefore appear in the normal mind. It was Dessoir's pioneering work that inspired his contemporary Norman Triplett to study this illusion empirically and later inspired me to study magic scientifically.

Dessoir's ideas on "negative hallucinations" are also of interest. The opposite of positive hallucinations, these refer to the seeing of nothing where something actually exists. Dessoir wrote, "Who has not hunted for an object that was directly before his eyes? The sense-impression exists, is taken up, not elaborate in consciousness and there in us arises a momentary state of mental blindness."[45] He went on to explain that "if we look at the hands of the magician and watch closely enough, we can see him conceal objects and exchange cards directly before the eyes of his audience. ... The exchange of cards, for example, falls within the observer's range of vision, the sensory irritation is made, but it does not come to consciousness."[46] This is a fundamental principle of misdirection. And a century later, psychologists (including myself) came up with experiments that proved him right.[47] What Dessoir observed by performing tricks for his fellow men of science is now known as *inattentional blindness*, an important focus of current research on visual perception.[48]

Dessoir also suggests that magic effects can contribute to the psychological understanding of free will:

> The well-known trick of permitting a card to be drawn at random, and immediately guessing it is based on the fact that the observer only believes he has freely dawn, while in reality the performer has restricted his will and diverted it in a definite direction, either by placing the card to be chosen in a convenient position, or by pushing it forward at the moment when the selector's fingers reach for it. I do not think that anything could offer a better illustration of the determinism of all our actions.[49]

This effect, forcing, has also recently become a subject of scientific interest.[50]

Around that time, Alfred Binet—a French psychologist who, among other things, invented the first intelligence test—also saw magic as a useful way to study the mind.[51] Binet invited some of the most famous magicians of his time to his laboratory in Paris, where he interviewed and tested them.[52] Binet was interested in learning from individuals with exceptional expertise, such as arithmetic prodigies. He thought that because magicians aimed to develop "conditions that can induce into us error and fool us as to what we are seeing," this group would be of particular interest. Binet went on to claim that "[one] can readily understand the interest of the psychologist in the study of the means employed to produce these illusions, for it enlightens us as to the process by which the mind perceives exterior objects, and makes known likewise the weak points of our knowledge."[53]

Binet took the scientific study of magic further by studying deception and misdirection using a revolutionary scientific device: the chronophotograph. After continually being fooled by the magicians in his lab, Binet obtained a chronophotograph, which let him take thirty photos per second at regular intervals. (It had previously been used to explore various human and animal activities and had provided, among other things, new insights into how cats turn right-side up when dropped.) Binet recognized that it is the mind, rather than the eyes, that are fooled—that magic exploits the way in which our brain interprets sensory information. Unlike our eyes, photographic plates do not interpret information, instead simply fixing the details of individual moments. Binet's magicians had their tricks photographed with the chronophotograph, capturing movements that lasted about a second (Figure 9.4). These short films

Figure 9.4

Raynaly drops a ball from his left hand into his right, performs a fake transfer of the ball back into his left hand, and finally opens his hand to show the ball has vanished. These images were digitized by Richard Wiseman, who managed to animate the images. Although they are only a few seconds long, they represent the earliest moving images of a magician. (Image courtesy of *Genii* magazine, from R. Wiseman, "The First Film of a Magician," *Genii* 69, no. 4 [April 2006]: 34–38.)

are the earliest magic movies. Binet found that removing the magician's commentary, misdirection, and fast movements destroyed the illusion. As such, this technique allowed psychologists to separate sensation from interpretation.

About the same time as Dessoir and Binet, Joseph Jastrow was exploring conjuring tricks.[54] He tested two prominent magicians (Harry Kellar and Alexander Herrmann) in his laboratory to determine their tactile sensitivity and reaction times. Although they were better than average on these measures, the differences were slight. One of Jastrow's contemporaries, Norman Triplett was inspired to publish an extensive article on the psychology of conjuring and deception.[55] As was fashionable at the time, he presented an categorization of conjuring tricks, as well as some general principles used in magic. Among other things, Triplett reported the first empirical studies on Dessoir's disappearing orange, which he called the Vanishing Ball Illusion (see chapter 5).

By the end of the nineteenth century, there was considerable scientific interest in the psychology of magic, but it ended shortly after. The next century witnessed very little further research on this topic, and much of what had been learned—as well as the approach in general—was forgotten.

It is impossible to know for sure why this happened. But a likely factor is that the new wave of psychology emerging at that time—behaviorism—had ruled out questions about internal experience. The psychology of magic, along with the study of attention and consciousness, simply did not fit into that way of thinking about the mind.[56] With the rise of cognitive science decades later (and its focus on internal mental structures), there once again emerged the scientific study of attention and later the scientific study of consciousness. It may now be time for a reemergence of the scientific study of magic.

A SCIENCE OF MAGIC: ATTEMPT #2

As a teenage boy, I had hoped to learn about psychology so that I could improve my magic. At the time, I read several popular psychology books on body language and psychological persuasion, as well as texts written by magicians. However, none of this was evidence based. During my university years, I was keen to learn more about the psychology of magic, but

I failed to find much empirical research about such phenomena. The link between psychology and magic seemed so obvious, yet there were so few scientific studies on this topic.

As I wrote my first scientific papers on misdirection in 2003, I was inspired by Peter Lamont and Richard Wiseman's book, which tried to link magic and psychology.[57] However, there was virtually no current psychological research on the topic. Most of my fellow psychologists liked the idea of studying magic but typically saw it as a bit of fun, rather than a serious science.

I was passionate about using magic to study the mind, so much so that after my PhD, I changed the direction of my research so that I could explore visual attention and misdirection. I was fortunate to gain a research fellowship at Durham University that gave me free reign to collaborate with researchers from all over the world to help establish a science of magic. Doing so was not easy and attracted some criticism. But I did manage to meet a few people who had similar ideas about magic and were able to help.

In 2008, Ron Rensink, Alym Amlani and I published a paper titled "Toward a Science of Magic," which set out our vision for this "new" field of research.[58] We argued that magicians had considerable knowledge of how to manipulate conscious experience and that their way of acquiring this knowledge had several key similarities with science. Most magicians have a theory about how each trick works, which they apply when designing new magic tricks. If their theory is wrong, the trick fails, and so they revise their theory and their trick, continuing to do so until they have a trick that works. This process is strikingly similar to the way that psychologists learn about the mind, except that psychologists test their theories using lab participants rather than live audiences. This informal process used by magicians has allowed them to acquire considerable knowledge about the human mind. We felt it was time for scientists and magicians to take this knowledge seriously and to develop the study of magic into a science.

Shortly after our paper was published, Stephen Macknik and Susana Martinez-Conde put forward a somewhat similar argument.[59] They teamed up with eminent magicians and tried to redescribe several magic tricks in terms of known neurological processes. Rather than psychology, they focused on neuroscience and later coined the term "neuromagic" to describe this field of study.[60] We were not convinced that we needed to restrict the science of magic solely to neuroscience. Although redescribing magic tricks in terms of neural mechanisms is important, on its own it is not enough to understand the nature of magical experience. Psychological and computational approaches are equally important and play critical roles in helping us gain a deeper understanding of magic's nature.

Meanwhile, a few academics were less than enthusiastic about the general idea of establishing a science of magic. For example, Peter Lamont and colleagues claimed that developing a science of magic had already been attempted and so was a "failed endeavor" that should not be attempted again.[61] However, one cannot dismiss a topic simply because earlier attempts to study it did not sustain; sometimes new things appear under the sun.[62] In the case of magic, technologies now let us measure behavior with far greater ease and precision than in the past. Instead of painstakingly recording sleight of hand with chronophotography, for example, we can simply use video or attach sensors to magicians' fingers to tell us the precise kinematics of their finger movements.[63] In misdirection research, we can employ eye-tracking technologies that let us measure where people look; these technologies are also providing new insights into related aspects of perception, such as visual attention.[64] Similarly, neuroimaging in the form of functional magnetic resonance imaging and electroencephalography allows us to pinpoint the neural mechanisms that are involved in experiencing magic tricks and other forms of deception.[65]

These technological advances are important, as are the advances in our understanding of the cognitive mechanisms involved. Many of these have been due to improvements in our understanding of how best to investigate these mechanisms. As a consequence, we have learned much about the brain and cognition in the last one hundred years, including new insights about several areas directly relevant to magic, such as attention, memory, and free will.

And finally—and perhaps just as importantly—changes have occurred in the culture of science itself. There is now a general recognition that internal experience is a legitimate topic of scientific study. The attitudes that dismissed the efforts of Max Dessoir and his colleagues a century ago are not found in mainstream research today. Things are now at the point where we can go beyond simply having an interest in magic—we can begin to study it scientifically and perhaps even develop it into a specialized area of science.

We believe that magic and science can meaningfully interact in at least three important ways.[66] First, the techniques that create magic can be used as tools to investigate scientific phenomena, independent of the experience of magic itself. Many psychologists have already done this to some extent without necessarily realizing it. For example, developmental psychologists frequently show babies simple tricks, such as making objects appear and disappear. Even though these tricks may not fool adults, children view them as magic.[67] And throughout this book, we have looked at various studies that use magic techniques to investigate a range of processes in adults. Misdirection is used to study attention, sleight of hand to create choice blindness, and magic apparatuses to duplicate cuddly toys.[68] Magic tricks have even been used to investigate the nature of problem solving.[69]

Second, the experiences produced by magic tricks are themselves of scientific value. For example, in chapter 3, we saw how magic can create anomalous experiences that scientists can then use to investigate why people endorse magical beliefs.[70] Likewise, the sense of wonder that people experience while watching magic is unique to magic and is of great scientific interest. As we have seen in the previous section, these experiences can also be harnessed to enhance well-being and education. As such, their natures and functions have important theoretical and practical implications.

Third, we believe that magic techniques themselves can also provide new insights into human cognition. For example, perceptual tricks used to hide objects are giving us new insights into perceptual processes.[71] Misdirection lets us examine how we orient our attention.[72] Forcing techniques inform us about the illusory experience of free will.[73] The "theory of false solution" allows us to gain intriguing insights into how our mind gets fixed on a solution.[74] Systematic investigation into magic tricks can even provide scientists with insights that are unique. For example, Cyril Thomas and I recently investigated an intriguing sleight of hand trick known as the Flustration Count, which gives you the impression of seeing different playing cards even though the magician is repeatedly showing you the same card.[75] It's a truly bizarre illusion that relies on a principle common in magic, but it also points to a form of perceptual reasoning not yet discovered by psychologists.

Ron Rensink and I have proposed that it may be time to consider developing an outright science of magic—a distinct area of study concerned with the experience of wonder that results from encountering an apparently impossible event.[76] This would be a science that has much in common with other sciences (e.g., visual science, neuroscience), with the aim of understanding and organizing what would otherwise be a rather disparate group of findings. We strongly believe that doing this will help us ask interesting and important new questions about the mind and help us further our understanding of cognition.

Peter Lamont has argued that there is too little structure in magic tricks for them to be studied in a systematic way, thereby ruling out a science of magic; Ron and I disagree.[77] The fact that there may be an endless number of different tricks does not prevent them from being studied scientifically. For example, there is a virtually infinite number of possible sentences, yet language can still be rigorously studied. Lamont also sees the lack of clear boundaries between individual tricks and performers as a problem; a trick carried out in a slightly different way is an entirely different entity. But this challenge has been faced—and met—in other sciences. Each individual animal is different (and can even change over time), but this does not impede biology. Such variations can be handled by carefully grouping together animals with largely similar characteristics. This kind of approach could be readily applied to magic tricks by carefully grouping together tricks that are experienced in largely similar ways.

I see no theoretically valid reason why a science of magic should be impossible. For example, my colleagues and I have proposed a psychologically based taxonomy to organize misdirection principles in a scientifically meaningful way, based on the perceptual and cognitive mechanisms involved.[78] I am sure there are errors in this taxonomy and that it will need future revisions. But it is a starting point that can help us bridge the gap between magic and science. I am also confident that the same principles can be applied to other aspects of magic as well.

Indeed, I feel rather optimistic about the prospects for this new science. To begin with, I am encouraged that so many researchers from a wide range of disciplines now use magic as a research tool. Magic has become accepted as a legitimate topic for scientific investigation; many university lecturers now teach students about it, and magic is even discussed in psychology textbooks.[79] These changes provide a springboard for students to carry out their own research, with our science of magic framework providing a road map for those interested in using magic to understand the human mind.[80] And along with this, the number of scientific papers published in this field has grown exponentially over the past decade. When I started teaching my course on the magic and the mind, the reading list was quite limited. There are now so many papers that few of my students can find the time to read them all.

Most importantly, a critical mass of enthusiastic researchers have now come together to help one another investigate magic. In the summer of 2017, we held the first science of magic conference at Goldsmiths, University of London, and as I am writing these lines, we are already planning our next conference. Over two days, more than one hundred scientists and magicians from all across the world came together to discuss science and magic and to hear about one another's research. I was amazed by the breadth, quality, and quantity of the scientific presentations given. There were also valuable insights from some of the world's top magicians (Juan Tamariz, Pit Hartling, and Thomas Fraps), who theorized and dissected the art of magic. It was a truly inspiring event that marked the reemergence of the science of magic as a legitimate scientific venture.

And in addition to all this, there is now a Science of Magic Association (SOMA), which promotes rigorous research on the nature, function, and underlying mechanisms of magic. Unlike the previous attempt by Dessoir and his contemporaries, there now seems to be sufficient momentum to keep the science of magic alive. In this book, I have used scientific research on magic to explain the cognitive processes that give rise to our experience of the world. Magic offers an engaging way to discuss psychology and provides tangible illustrations of our cognitive processes. Even more importantly, it can help uncover new aspects of cognition. Only time will tell how far and in what directions this venture will ultimately take us.

HOW TO ADVANCE THE MAGIC ENDEAVOR?

BACK IN THE 1870S, Jean-Eugène Robert-Houdin wrote,

> The art of conjuring bases its deceptions upon manual dexterity, mental subtleties, and the surprising results which are produced by the sciences. The physical sciences—generally chemistry, mathematics, and particularly mechanics, electricity, and magnetism—supply potent weapons for the use of the magician. In order to be a first-class conjuror, it is necessary, if not to have studied all these sciences thoroughly, at least to have acquired a general knowledge of them, and to be able to apply some few of their principles as the occasion may arise.[1]

Deception is central to magic, but good magic involves more than merely fooling people. Magic needs to astonish and enchant, and the performance needs to engage and entertain the audience. Magicians often take inspiration from other areas, such as film, theater, literature, comedy, dance, and even science. Robert-Houdin's advice was written at a time when psychology was still in its infancy, but I am sure that if he were alive today, psychology would be added to his list of sciences. Indeed, as Juan Tamariz explains,

> Psychology is, without a doubt, the most essential aspect, and without it, it is practically impossible to be a good magician. I am referring to the knowledge (intuitive and acquired) of the psychological mechanisms of the minds of the spectators, such as knowing in detail what "blind spots" are present in the spectator's mechanisms of perception, attention, and memory. This psychology also allows us to know when it is possible to achieve an illusion within their senses, and make them perceive things that are not really happening.[2]

I started studying psychology because I wanted to improve my understanding of conjuring, and there were few magicians who doubted the fundamental role of psychology in magic. However, the idea that science can improve magic is more controversial. I have given several lectures at the Magic Circle, during which I ran an informal survey to assess my fellow magicians' views. To my astonishment, nearly three-quarters of the audience thought that science could not help improve magic (although some have since changed their minds). We recently held a panel discussion on this topic at the last Science of Magic Association Conference. Before taking my seat, I was worried that the discussion might be unnecessary—surely all of these attendees would agree that science can add value to magic. I was wrong. Most magicians believe that because they have been experimenting with psychological principles for nearly two millennia, magicians have a substantial head start on psychologists. Psychologists, therefore, can scarcely provide new insights into their art form.

As a scientist who tries to bridge the gap between science and art, I am often challenged by both sides. Scientists prefer well-controlled experiments and often shy away from noisy real-world contexts. I personally enjoy the challenge, and doing so has allowed us to uncover intriguing new phenomena that would have been missed if we were confined to tightly controlled laboratory settings.[3] Likewise, magicians often perceive themselves as creative artists and are, perhaps understandably, rather reluctant to take guidance from scientists. I have been privileged to chat with some of the greatest magicians and scientists, and both groups include some of the brightest and most creative individuals you will ever meet. Magicians and scientists are surprisingly similar, so why has it been so hard to bridge this gap between magic and science? Throughout history, magicians have always absorbed cutting-edge knowledge from a wide range of experts. Why are they so reluctant to take advice from psychologists, a group of individuals dedicated to understanding the mechanisms that underpin our thoughts and behaviors?

I believe this reluctance originates from some individuals' perceptions of what science is and aims to do. The magicians in Alfred Binet's studies probably did not learn much from the chronophotography analysis, nor have any of the magicians who participated in our own studies.[4] Similarly, I am doubtful that knowledge about the neural mechanisms that underpin the experience of magic will help improve magic tricks.[5] I also understand why magicians might be disappointed by some of the science of magic research, which is intended to help understand cognitive processes rather than improve magic. As a scientist, I strive to understand the mind rather than help magicians, but this does not preclude the use of science to enhance the art of magic. More generally, I would like to propose a more systematic and analytical approach to magic, and in this final chapter, I will highlight how and why this approach might advance the magic endeavor.

IMPROVING MAGIC METHODS THROUGH SCIENCE

Magicians are true masters of deception, and their experience in performing for large audiences gives them amazing insights into how their techniques work. Magicians and psychologists have much in common: they both try to find errors in our cognitive processes. Richard Wiseman claims that magicians are much better equipped to understand the mind because their effects are more powerful than those typically studied in the lab.[6] Whereas psychologists are happy for an effect to work 80 percent of the times, a magic trick that only fools 80 percent of the audience is simply not good enough. The last point is certainly true, and much of my motivation for developing a science of magic rests on the idea that magicians have valuable insights into our mind. Yet this does not preclude the usefulness of more rigorous scientific methods. Magicians are knowledgeable about what principles work, but they do not necessarily understand why they work. I believe that science can improve magic in two important ways. First, a scientific understanding of the mental processes that underpin our experiences allows magicians to create more powerful tricks. Second, new scientific tools allow us to evaluate magic theories more objectively, which will result in more evidence-based knowledge. Let us now take a detailed look at how this can be achieved.

The human mind is extremely complex, and even though there are many aspects that we do not fully understand, advances in psychology and neuroscience can provide valuable insights into human cognition. We are now in a position where scientists can explain, or at least speculate, about why some magic tricks work. Most illusion designers would happily take advice from a structural engineer, so why turn down advice from brain scientists who devote their lives to studying the mind?

Decades of scientific research have provided insights into why our perceptual experiences are often removed from reality. Understanding these mechanisms would allow magicians to create more powerful illusions. Let me give you a simple example. Change blindness and inattentional blindness are scientific discoveries that took most scientists by surprise and changed our theories about consciousness. These perceptual failures highlight gaps and blind spots in our conscious experience that few of us had imagined. Although magicians exploit many of these perceptual failures, I do not think they fully appreciate their magnitude, nor do they fully understand why these limitations occur. For example, in my lecture at the Magic Circle, I included several change blindness demonstrations that took many of the magicians by surprise, and like most people, they were vulnerable to overestimating how much people truly perceive.

Many magic tricks rely on change blindness, yet it was visual scientists who developed clever experiments that allowed them to discover the phenomenon in the 1990s.[7] These scientific findings have inspired several tricks and illusions. For example, Richard Wiseman has designed a wonderful color-changing card trick that is based on change blindness.[8] Other magicians, including Penn & Teller and Tom Stone, have also created tricks inspired by change blindness.[9] However, I believe that we can take this even further.

Inasmuch as all of us (including magicians) intuitively overestimate the amount that we consciously perceive, magicians could be developing bolder and more daring techniques. Magicians typically assume that attention simply refers to where you look, but our work shows that people often miss seeing things that are right in front of their eyes.[10] In one of our experiments, we used a color-changing card trick in which we deployed social misdirection (i.e., asking someone a question) to prevent people from noticing that the back of the cards had changed from blue to red. To our surprise, people were just as likely to miss the color change when they were looking at the cards as when they looked at my face.[11] This study was not intended to help magicians, but the findings can help create more powerful effects. For example, a trick in which a card changes while the spectator keeps his eyes on it would be pretty spectacular—and our findings suggest that this is possible. We now have pretty good scientific models of how attention works, and these models help predict the conditions under which we notice things. For example, the exact nature of a task (e.g., instructing people to focus on the team in white rather than the team in black) changes the chances of spotting the gorilla.[12] Similarly, the information that you hold in working memory influences what you attend to and the ease with which you are distracted.[13] Even though magicians have much real-world experience in manipulating people's attention, understanding how attention operates will help fine-tune attentional misdirection.

There are many other areas in which scientific knowledge can improve magic. For example, understanding the psychological factors that lead to memory errors (e.g., source-monitoring errors) will help ensure that the audience misremembers important details of a trick. Theories on agency and free will help explain why people genuinely believe a forced choice to be freely made. I always thought it was much better to ask someone to think of a card rather than physically pick one, yet the scientific study on forcing and agency suggests that both are experienced as being equally free.[14] Again, understanding the psychological mechanisms that underpin our sense of agency will help develop more powerful forcing techniques. There are countless other examples, but this might be a topic for another book. However, let me just conclude that even though magicians may have superior experience about which tricks work, science helps explain why they work. This knowledge can help improve existing methods and possibly design new ones.

Scientists ask questions and use data to evaluate them. When I have a budding new theory about how the mind works, I try to come up with an experiment to test it. In psychology, we often manipulate certain factors and then establish whether this manipulation changes people's behavior. The challenge for most psychologists lies in developing suitable manipulations and using accurate, as well as meaningful, measurements. Doing so is far from easy, which is why scientists spend ages quarreling. Magicians also like to debate their theories, and although they often try to back them up with data, these data typically take the form of anecdotal evidence.

Measuring the effectiveness of a magic trick is difficult, which is another area where science can help. Most magicians use informal observations, based on how people react and applaud, to evaluate whether a trick worked.[15] Talking to people after the performance also provides useful feedback. For example, people often spontaneously tell me about the things they enjoyed and sometimes speculate about how the trick may have been done. This information gathering is useful, but psychology has taught us that informal observations are often biased and unreliable. For example, people may simply be too polite to tell me that my trick was rubbish, and there are lots of psychological factors that can influence the level of applause, making it a rather unreliable measure of enjoyment. As we have seen throughout this book, our intuition about our own and other people's experience is often wrong. And if you truly want to evaluate a magic trick, you need more objective measurements.

Objectively evaluating a magic trick is far from easy. I used to practice in front of the mirror, and today's recording equipment allows me to film my performance and review it in more detail. Despite the fact that this feedback is useful, my own experience of the trick does not reflect how others will perceive it. For example, a recent neuroimaging study has shown that seeing the same trick led to different neural activation in the performer compared to the naïve audience.[16] Asking my magician friends for help is also difficult, because they often know the secret and bring along a very different set of assumptions, which changes their experience of the magic. As a teenager, I often tried my tricks on my little sister, but after a while, she too became an expert, and her opinion no longer reflected the way that people typically perceive magic.

The most straightforward way to evaluate a magic trick is to ask members of the audience directly, but this is also challenging. Magic relies on people's misconstruction of the trick, and it is often difficult to question someone without giving away parts of the secret. For example, when forcing a card, few magicians will quiz the spectator about whether their choice was truly free. Similarly, I have never asked anyone about whether they saw me load the lemons under the cup or whether they notice the bulge in my coat, because doing so would reveal part of the secret and destroy the illusion. Moreover, the role of the magician is to entertain, and most audiences will not necessarily welcome a postperformance interrogation/interview/pop quiz. However, shying away from these important questions will prevent us from truly discovering why magic works and thus prevent us from improving it.

Psychologists develop tools that allow them to measure a wide range of behaviors. For example, advertisers typically use focus groups to evaluate different versions of an advert before a campaign goes live, and many film production companies use test audiences to evaluate different versions of a film. These psychometric tools can also be used to evaluate magic tricks, and some magicians have started to do so.[17] By performing in front of focus groups, we can evaluate different versions of a magic trick and ask questions that would simply be impossible otherwise, which will give us a deeper understanding of the method.

Joshua Jay took psychometric testing a step further by teaming up with psychologists who developed surveys to test what people really think about magic.[18] This was the first scientific study intended to help magicians, and the results were instructive. It identified the aspects of magic that people truly like (e.g., surprise) and the type of tricks deemed most enjoyable (mentalism). This survey contained vast amounts of data about what people truly think about magic and provided valuable objective information that will help magicians improve. Surveys and focus groups can never provide a complete picture of a magic performance, but these more objective approaches will provide much more accurate feedback and thus advance the art of magic.

Understanding why things work often involves manipulating psychological parameters and discovering conditions that change behavior. Magicians regularly use trial and error, but they often stop experimenting once a trick works. As a scientist promoting the science of magic, I shamefully admit that I rarely change the parameters of my Cups and Balls routine. Why change it when it works? For example, although I have spent many hours debating the importance of using real fruit for the final reveal, I have never tested this assumption empirically. I am also reluctant to manipulate my misdirection, because I do not want people to see how I sneak the lemons under the cups. Magicians are paid to entertain, and few can afford to reveal their secrets to further our understanding of why the tricks work. As a scientist, I am in an entirely different position, and it has been fascinating to change magic tricks so that they no longer work.

Trial and error certainly advance our understanding of magic, but scientific approaches are more effective. I recently came across a blog post that beautifully describes a scientific approach developed by a magician.[19] The anonymous magician tested different versions of a magic trick in front of focus groups. Unlike typical audiences who pay to see magic, these individuals were paid to watch the tricks, which meant that the magician was free to extensively quiz them about their thoughts and experience. It also meant that he could change parts of his tricks without having to worry about whether the audience was being entertained or whether they would figure out how the tricks were done. Indeed, finding conditions that break an illusion are essential when trying to figure out why magic works. These systematic investigations were conducted without fancy lab equipment, and they illustrate how the scientific method can be applied to test magic theories. Applying this approach more broadly will provide a deeper understanding of magic and ultimately allow us to develop more powerful techniques.

Magicians often use computer technology in their tricks, and there are countless electronic devices and smartphone apps that can be used to create new tricks. However, Peter McOwan and Howard Williams (whom I met when he started his PhD) have taken computer science and magic a step further: they use artificial intelligence to optimize existing tricks and even create new ones.[20] I was skeptical at first, but I am now convinced that this revolutionary approach has lots of potential. As the science of magic helps identify more psychological parameters that underpin the tricks, we can use those parameters to come up with the perfect trick. However, most tricks rely on lots of different parameters, each of which can be manipulated independently, thus resulting in a vast number of combinations. It is therefore unfeasible, if not impossible, to physically try out all the different combinations. Modern computers, however, can carry out vast amounts of calculations in a fraction of the time it would take us to do them manually, and thus computers provide the perfect tool to test different parameter combinations. Computational approaches use clever algorithms to create and test a vast number of new prototypes and help select the best combination. This approach has been applied to a wide range of creative domains (e.g., design, music, poetry) and is now being applied to magic—as illustrated by Howard's trick "The Twelve Magicians of Osiris."[21] Howard isn't a magician, and the trick won't win him a prize at the next world championship, but it is a powerful demonstration of how computational algorithms can find the perfect parameter combination to create the most powerful effect. I am confident that this approach could also be used to enhance the effectiveness of several stage illusions and of magic in general.

This computational approach can also be applied to create some rather powerful mind-reading effects. A previous study by Jay Olson and colleagues showed that people prefer some cards (e.g., the queen of hearts) over others.[22] Williams and McOwan used their computational approach to develop a stacked deck of playing cards in which any four cards dealt from the deck will result in just one liked card being dispensed. If the spectator is asked to look at these four cards and simply think of the card that they like most, there is a high probability that they will chose the target card—a new and rather powerful card force. Williams and McOwan have used artificial intelligence to create numerous other magic techniques, and their approach provides an inspiring glimpse into how science can help improve old tricks and even create new ones.

WHY DO WE ENJOY MAGIC?

Magicians love to talk about conjuring techniques, but understanding why we enjoy magic is equally important. Although there are thousands of books and articles that discuss how to perform a trick, they rarely examine why people enjoy being tricked and what distinguishes it from merely feeling puzzled. Answering this neglected question is not easy, but advances in psychology are providing new insights with important ramifications for our understanding of magic. There are lots of reasons why we enjoy magic, and the enjoyment might vary between individuals and the tricks themselves. Magic encompasses a wide range of styles and genres, making it difficult to generalize across them all, and there has been remarkably little research on this. However, we can compare magic to other performance arts and focus on the feature that is unique to magic (i.e., experiencing the impossible), which provides a starting point to explore the psychological mechanisms responsible for our enjoyment of magic.

Magic creates a conflict between the things we experience and the things we believe to be possible, and it can elicit a wide range of emotional responses: joy, amazement, wonder, surprise, vulnerability, loss of control, apprehension, fear, interest, curiosity, confusion, and bafflement.[23] Although it is easy to explain our enjoyment of the positive emotions, the negative components are more challenging. For example, Jason Leddington suggests that we experience a magic trick as an aporia, a form of intellectual bafflement and cognitive failure.[24] You know it is a trick, yet you simply cannot explain how it's done. From this perspective, magic should elicit distinct negative reactions to the apparent violation of our understanding of the world and to our failure to explain this violation (i.e., not knowing how the trick is done). These negative emotional experiences are central to any magic trick, yet we still experience it as aesthetically pleasing and entertaining.

One way to explain this paradox is to suggest that the negative emotions are compensated for or healed by positive antidotes, such as comedy or some other spectacle. Jamy Ian Swiss suggests that because it is "not fun to be fooled ... [i]t is one of the magician's first orders of business ... to add a spoonful of sugar to help the medicine of magic go down."[25] It is true that magicians often incorporate comedy into their performance, but Leddington points out a fundamental flaw in this argument.[26] If our enjoyment of magic relies on humor to compensate for the negative experiences, why do people enjoy the magic itself? Is it not easier to drop the magic and simply keep the surrounding theater? Moreover, there are lots of instances where you enjoy the magic without the humor. In fact, Darwin Ortiz suggests that humor can often be distracting, and some of the most profound experiences result from simply observing strong magic without the theater.[27] Even though there is clearly an interesting connection between humor and magic, we can only understand magic by making sense of these negative emotions.

The seeming paradox of embracing both positive and negative emotions is not unique to magic, and recent research on the psychology of aesthetics may provide some answers to this paradox. Many artistic genres (e.g., dramas, novel, film, paintings, poems) elicit a mix of emotions, such as fear, sadness, or even disgust. In most genres, our pleasure actually relies on these negative emotions.[28] The most striking example is our enjoyment of horror movies, where people genuinely enjoy the negative affect that these films elicit. Horror aficionados positively embrace the fearful feelings themselves, rather than the emotional antidotes (e.g., moments of relief, happy endings).[29]

A recent psychological theory on aesthetics explains how and why people enjoy these negative emotions.[30] Negative emotions have desirable psychological properties: they effectively capture your attention and lead to intense emotional involvement that typically results in highly memorable experiences—potent qualities for any art form to achieve. The *distancing–embracing theory* proposes that two different psychological mechanisms are activated as you indulge in an artistic experience.[31] First, a distancing mechanism allows you to distance yourself from the experience, providing you with personal safety and control over whether or not you want to continue. This process prevents the negative emotion from becoming incompatible with your expectations about what is enjoyable. For example, as you are watching a magic show or a film, your situational awareness tells you that what you are seeing is not real. Few would enjoy seeing a person being mutilated for real, but as you are watching a horror film, your "art schema" allows you to distance yourself from the context. Likewise, experiencing violations of your world knowledge in a real-life context could be rather frightening and unsettling, but in the context of the magic show, you can fully embrace these negative emotions (e.g., loss of control, aparia). These distancing mechanisms set the stage for a second psychological mechanism that allows you to embrace negative emotions, thus potentially making art seem more intense, interesting, emotionally moving, and profound. Most importantly, it means that magic can embrace a wide range of emotions (both positive and negative) that will ultimately lead to a richer experience.

The distancing–embracing model also explains some other interesting psychological properties. For example, the negative emotions explain why magic performances are often captivating, emotionally powerful, and memorable.[32] Understanding these mechanisms gives us a deeper appreciation of this art form and can help create more powerful experiences. I love card tricks, which are one of the most popular genres among magicians. However, people often forget these tricks—ironic considering their often surprisingly vivid memories of other tricks, sometimes seen several decades ago.[33] For example, I am often surprised that people remember some of the tricks I performed several decades ago and recount magic performances they saw as a child. These long-lasting memories perhaps result from the powerful mixtures of emotions that magic elicits, whereas a bland card trick may simply fail to activate these emotions. The distancing–embracing model tells us that we should fully embrace an experience if it elicits a wide range of powerful positive and negative emotions. As long as people can distance themselves from the performance, it will lead to profound and truly enjoyable experiences. It comes as no surprise that some of the world's most successful magicians, such as Penn & Teller, use shock, disgust, and fear to create truly enjoyable and legendary performances.

There are other reasons why people are captivated by magic. For example, we are intrinsically motivated to understand the world, and we are probably drawn to magic because of a deep-rooted drive to explore things we do not understand. From an early age, infants are captivated by events that violate their understanding of the world, such as when objects defy the laws of physics, and the same is true for adults. People are intrigued by things that don't make sense, and recent research has uncovered interesting systematic differences in our interest in such events.[34] Thomas Griffiths asked participants to rate their interest in a wide range of magical transformations. The resulting data showed that people were more interested in transformations that moved in an ontological hierarchy (i.e., between different categories), such as the direction of animacy as opposed to the other way around. For example, their results showed that people are more interested in a vase that transforms into a rose than a rose transforming into a vase. Some of these principles are implicitly implemented in magic, but a full understanding of the cognitive mechanisms that underpin these variations can help magicians create more interesting and engaging tricks.

Magic provides a glimpse into a world that is not restricted by the everyday laws of reality. People immerse themselves in fiction and fantasy because it provides an escape from our mundane everyday reality.[35] Magic pushes the boundaries of what we believe to be possible and provides a compelling illustration of an enchanting alternative reality. Although there is a fundamental difference between magical fantasy and stage magic, similar psychological mechanisms may explain their appeal. We have already seen how magic tricks can change children's beliefs in magic, and in adults, the distinction between real magic and stage magic is often rather blurred.[36] Adults often struggle to distinguish between trickery and real magic. For example, in some of our experiments, numerous participants who were explicitly told that they were watching a magic performance still believed it to be a genuine psychic demonstration.[37] We also asked our participants to rate how much they enjoyed the experience, and there was a positive relationship between enjoyment and the extent to which they believed it was genuine. Mentalism is a magic genre that often blurs the boundaries between magic and reality, and despite the fact that people rarely believe in my magical powers when I vanish a ball, they are often more willing to entertain the belief that mentalism is real. This might explain why mentalism is the most popular magic genre.[38] However, for all strong magic, there is a split second in which the audience believes that what they are seeing is real, and I believe that this brief glimpse into a magical world is truly enchanting and enjoyable.

We are still in the early stages of understanding people's experience of magic, and much more empirical and theoretical work is required to fully explain our enjoyment of magic. However, I am convinced that a deeper psychological insight can help create more powerful and enjoyable experiences. There are still many unanswered questions. For example, why do some people enjoy magic while others don't? Why do some people strive to discover how the trick is done, yet others are happy to accept the mystery? Although most magicians report large individual differences in how magic is received, we have so far struggled to find any obvious personality traits (e.g., openness to experience) that predict an individual's enjoyment for magic. However, I am confident that such traits will be found in future. For example, Joshua Jay's survey found that women generally prefer magic more than men do, but we do not know why.[39] I believe that understanding these differences can help magicians tailor their performance to different audiences.

THE PERFORMER: MAGIC AND GENDER

Our enjoyment of magic relies just as much on the person performing the trick as it does on the trick itself. Any successful endeavor relies on attracting bright new talent, and most organizations in sports, arts, and sciences actively try to recruit newcomers. The same is true for magic. Understanding why some individuals dedicate large parts of their lives to deception—as well as the barrier that prevent others from taking up magic as a hobby or profession—will help enrich the magic community and help advance the magic endeavor.

Unlike most other forms of art and entertainment, however, there is no formal magic training, and I'm not aware of any degrees in magic. Indeed, we know relatively little about the magic community, but my PhD student Olli Rissanen has interviewed a group of professional magicians to find out more.[40] Olli looked at the skills and professional competencies that successful magicians require, and he also asked them why they got into magic. Most magicians claimed that they were inspired by a memorable magic performance that they saw as a child and that this experience motivated them to take up magic as a hobby. Although I am sure that these childhood experiences can have profound effects, most children are enchanted by magic and receive magic sets, yet few ever take up magic as a serious hobby.

There is one psychological factor that predicts more than any other whether you take up magic as a hobby: gender. Less than 10 percent of all professional magicians are women, and even though there has been a visible and welcome increase in the number of female magicians, it is one of the most gender-divided forms of entertainment, closely rivalled only by comedians.[41]

In history, female magicians have been associated with the dark arts or witchcraft, and it is often stated that throughout the fifteenth, sixteenth, and seventeenth centuries, woman who dared to preform magic tricks were considered an incalculable threat and consequently were drowned or burned at the stake.[42] This threat was certainly off putting, yet there is no actual evidence that entertainment magicians were ever tried for witchcraft.[43] By the eighteenth century, magic had moved from the street into the theater, and the illusions became more elaborate and often required the help of assistants, who were typically children or women. In the nineteenth century, magic became more popular, and women found success in roles as psychics and mediums.[44] There were, however, some female magicians at the time. For example, Adelaide Herrmann was married to one of the most popular American magicians, and after his death in 1896, she took over his show and toured the globe for nearly thirty years, making her one of the most famous female magicians.

Historians often overlook female magicians, and unfortunately, they receive very little attention in most magic books. Throughout the magic literature, magicians are typically referred to as men, unless they act as an assistant.[45] For example, one of the best-known books on magic from the 1960s devotes an entire chapter to "making the most of assistants," and the author continuously refers to these assistants solely as women.[46] The author even suggests that the magician find the right "girl" to fit his ideal characterization. Female assistants often endured excruciating physical demands and the risk of injury or death. Although these women carried out 90 percent of the work, they were encouraged to simply fade into the background and rarely got the respect and admirations that their male counterparts did.[47] Such role inequality was sadly not unusual throughout history, and the imbalance has persisted into recent times.

There are several sociological and social-psychological factors that can potentially explain this gender imbalance. Peter Nardi suggests that the social organization of magic and the social structure of magic clubs help foster the male dominance in magic.[48] For example, one of the world's leading magic clubs, the Magic Circle, was an exclusively male society until it opened its doors to women in the early 1990s. Traditionally, these organizations acted as gatekeepers, which made it more difficult for women to access mentors and professional contacts. With advances in social media, the role of these traditional gatekeepers is fading, which may help women get into magic more easily.[49]

Performing magic involves a kind of deceit that involves power, control, and one-upmanship. Nardi suggests that "magic is an aggressive, competitive form involving challenges and winning at the expense of others."[50] He suggests that because men are more frequently encouraged to demonstrate power, control, and competitive manipulation of others, magic is more compatible with the skill set fostered in males. It is also likely that the male image promoted in the magic community and the lack of female role models contributes to why so few girls take up magic as a hobby.[51]

Most magic books have been written by men, for men, and many performance traditions are simply not designed with a female performer in mind. For example, many tricks assume that you are wearing a jacket with long sleeves and a pair of trousers with pockets. This means that I can put my hand in my pocket to get rid of a secret prop and that I also have sufficient space for my lemons. Clothes play an important role in magic, and I spent much of the 1990s wearing unfashionable blazers simply because they allowed me to perform certain tricks. Even today, I often agonize about finding ways to justify wearing a coat during my lectures so that I can perform the Cups and Balls. Female fashion is less versatile for magic, which makes it difficult to perform certain tricks. Jeniffer Ortega is one of my female PhD students who is working on misdirection and also performs magic in her spare time. A few months ago, we were talking about a card trick, and I advised her to simple switch the normal cards for a gimmicked deck. She looked at me half bemused and explained that as she rarely wears a coat with pockets, this was much more difficult for her.

It is not just the methods that have been designed for men but also the patter (i.e., what magicians say). Most good magicians write their own scripts, but for novices, it is often easier to use tried and tested patter. For example, most magic books and video tutorials provide detailed scripts, which novice magicians often copy verbatim. Magicians are notorious for stealing one another's jokes and funny lines, and their presentation styles are often rather gendered. It is relatively easy for male magicians to simply copy another man's performance style. Women, on the other hand, must be more resourceful and creative, and they need to tailor the presentation to their own gender.[52] This is not problematic in itself and, in the long term, can help promote a more unique performance style, but it does add an additional challenge for novice female performers.

Women face a lot of additional hurdles in performing magic, and we have recently discovered that simply knowing that a trick is being performed by a female magician changes the way that people perceive it. In a research project lead by Pascal Gygax, participants watched short videos clips of magic tricks and were told either that it was performed by Natalie (a female magician) or that it was performed by Nicholas (a male magician).[53] Although all of the tricks were performed by the same magician, people thought the "male" performer was better than his "female" equivalent. It is astonishing, but possibly not surprising, that female magicians have to work significantly harder to impress their audience than their male colleagues.

Magic is one of the most gender-imbalanced forms of entertainment, and it is important to learn about the sociological and psychological challenges that female magicians face. Doing so will hopefully help women gain a more prominent status within magic, help bring diversity to the community, and, I believe, enhance the art of magic as a whole.

SUMMARY

Throughout history, magicians have taken inspiration from a wide range of disciplines, and there has been a particularly close link between magic and science. Victorian magicians prided themselves on being men of science.[54] Magicians are extremely creative, and their real-world experience in deception provides them with unique insights into their art form. I believe that combining this real-world knowledge with scientific insights into why the tricks work will help advance the art of magic. Moreover, the scientific methods used to help understand the human mind would help us gain a more objective and deeper understanding of magic. Likewise, a more scientific and analytical approach to magic would help ask questions that magicians often ignore. Although magicians love to know how to fool people, they rarely ask why people enjoy being tricked. Psychological theories allow us to answer some of these questions, and understanding the nature of this experience can help create more profound magic tricks. A magic endeavor will only be as good as the individuals involved in it, and as such, it is essential to understand the barriers that prevent individuals from participating. In this chapter, I have focused on the gender imbalance and some of the factors that prevent women from participating in magic. Understanding the additional challenges faced by women and other marginalized individuals (e.g., ethnic minorities) will help create a more diverse and richer magic community, which I believe will enhance the magic endeavor. I use magic to gain new insights into the human mind and believe that this is an extremely valuable approach to science. Likewise, I believe that a more systematic and scientific approach to magic will help magicians gain a deeper understanding of their art from and ultimately advance the magic endeavor.

$\textcircled{11}$

CONCLUSION

CONJURING IS ONE OF THE OLDEST and most enduring forms of entertainment, and magic has played an important role in many aspects of our lives. Even though the Enlightenment brought scientific explanations for most of the ancient mysteries, advances in science and technology have done little to squash magic's appeal. To the contrary, magic has rarely been as popular as it is today.

Magicians amaze and enchant, but there is more to magic than simple entertainment. Magic deals with some of the most fundamental philosophical and psychological questions (e.g., consciousness, deception, beliefs, free will), yet it has received far less academic attention than other art forms, such as music, film, literature, and the fine arts. Despite the fact that thousands of magic books have been written for magicians (by magicians), very few are intended for outsiders. Ignoring this fascinating form of entertainment is a lost opportunity. The past decade has seen huge advances in our understanding of magic, yet there is still much catching up to be done.

Magic is shrouded in clouds of secrecy and deception, which often prevent outsiders from studying it. Some magicians are reluctant to reveal their secrets for fear of destroying their illusions. Magicians also often mislead outsiders about how their effects are achieved. Throughout history, magicians have presented their tricks as scientific demonstrations, and today, they often use pseudoscientific neuro-jargon to create intriguing illusions. For example, some magicians use tricks to demonstrate psychological skills, such as priming or body language reading, and people often take these explanation at face value. However, as we have seen throughout this book, there is often more to magic than meets the eye.

Magic relies on deception and secrecy, but this should not prevent us from studying it more systematically. Most magicians are pretty open about their deceptive techniques, and the idea that they guard their secrets with their lives is an illusion. I have tried my best to avoid revealing secret methods unless necessary, but it is impossible to please all. For the non-magician eager to know how the tricks are done, I simply say: the secrets to most magic tricks can be unlocked with a few clicks on the internet. But be warned. Once you know the secret, you will never again experience the wonder and mystery.

Magicians are often concerned with exposure, and theoretically, I could be expelled from the Magic Circle for revealing how magic tricks are done in public. However, magicians are allowed to sell their secrets, and inasmuch as you have all paid to buy this book, there is little danger of this. In the digital age, it is virtually impossible to keep anything secret, and thus social media platforms and YouTube pose a big challenge to magic.[1] These open platforms, however, also provide new opportunities for the magic endeavor. I learned magic from books that I found in our local library, while today's generation is learning from watching YouTube tutorials performed by masters. Young magicians today learn new and better techniques, and I am often amazed—and put to shame—by their skills. Magicians should embrace the challenge of openness because it will ultimately encourage them to develop better tricks.

I believe that a more open approach to magic, one that allows outsiders to study and contribute, will enrich the endeavor. For any magicians concerned about exposure, rest assured that most people don't actively seek out the secret; they enjoy and cherish the wonder too much to do so. I hope to have offered some insights into this wonderful form of entertainment without destroying the wonder that it elicits. I hope this will help enhance the magic endeavor and help people appreciate this often neglected art form. Most importantly, though, magic offers many important benefits beyond simple entertainment.

Magic provides compelling illustrations of our mind's limitations, and it highlights the subjective nature of our conscious experience. The science of magic offers a new perspective on cognition, and it provides compelling illustrations of the surprising tricks our brain uses to deal with the challenge of survival. We rarely reflect on these challenges, and our cognitive processes are often impenetrable to conscious introspection. Most importantly, magic creates a conflict between the things that we believe to be possible and the things that we experience, and it does so by exploiting our delusional beliefs about our own cognitive abilities. Understanding and appreciating these errors in meta-awareness has far-reaching implications for how we judge ourselves and others. Moreover, been fooled by magic is a compelling reminder of how much we misjudge our own capability.

These errors and illusions provide scientists with new insights into the brain, but they also help us make more informed decisions. Technological advances are changing the demands our brains face, and our safety relies on realizing our own limitations. For example, the ease with which magicians can create blind spots by manipulating your attention explains why it's a bad idea to talk on the phone while driving. Similarly, many of the perceptual and memory illusions that we explored in this book challenge our thoughts about the way we perceive the present and remember the past. The ease with which magicians influence your decisions shows that the compelling sense of free will we typically experience may be an illusion. Throughout modern history, people have tried to persuade others and influence public opinion; advertisers try to convince us to buy their products, and politicians try to manipulate public attitudes. Democracy relies on informed and free choices, yet in our digital age, the distinction between "real" and "fake" news is becoming more blurred, and it has become much easier to covertly manipulate opinion. As with the magician's force, we feel like our choice is free, and yet this may simply be an illusion.

The science of magic provides valuable lessons. It is difficult, if not impossible, to enhance our cognitive resources, and these forms of covert persuasion are often hard, if not impossible, to resist. However, awareness of these limitations and biases has the potential to help us make more informed decisions and to reduce the chances of us being unknowingly manipulated. Although magic highlights many of our limitations, we should also embrace the positive experiences that magic elicits: wonder, creativity, captivation. A deeper understanding of the nature of magic will help translate these principles from the stage to our daily lives. So much more can be learned from magic and the psychological mechanisms underpinning these experiences, and we hope to see significant advances in this area in the years to come.

NOTES

CHAPTER 1: WHAT IS MAGIC?

1. For example, see R. A. Rensink and G. Kuhn, "A Framework for Using Magic to Study the Mind," *Frontiers in Psychology*, February 2, 2015.

2. See S. L. Macknik, S. Martinez-Conde, and S. Blakeslee, *Sleights of Mind: What the Neuroscience of Magic Reveals about Our Everyday Deceptions* (New York: Henry Holt, 2010).

3. J. Jay, "What Do Audiences Really Think?," *Magic*, September 2016, 46–55.

4. See G. M. Jones, *Trade of the Tricks: Inside the Magician's Craft* (Berkeley: University of California Press, 2011). Julia Misersky and her colleagues asked people across several different cultures to rate 420 different role names and professions according to whether they were more typically carried out by men or women. Magicians scored in the top 18 percent of male professions, exceeded only by hunters, presidents, and terrorists. It is not my intention to promote this gender stereotype, and we will discuss the role of gender in magic later on. However, for simplicity, I will refer to the magician as "him." J. Misersky, P. M. Gygax, P. Canal, U. Gabriel, A. Garnham, F. Braun, T. Chiarini, et al., "Norms on the Gender Perception of Role Nouns in Czech, English, French, German, Italian, Norwegian, and Slovak," *Behavior Research Methods* 46, no. 3 (September 2014): 841–871.

5. O. Rissanen, T. Palonen, P. Pitkänen, G. Kuhn, and K. Hakkarainen, "Personal Social Networks and the Cultivation of Expertise in Magic: An Interview Study," *Vocations and Learning* 6, no. 3 (October 2013): 347–365.

6. See J. Tamariz, *The Magic Way* (Madrid: Editorial Frakson Magic Books, 1988).

7. See R. P. Wilson, dir., *Our Magic* (Los Angeles: Dan and Dave, 2014), DVD.

8. See also D. Ortiz, *Designing Miracles* (Oakland, CA: Magic Limited, 2006).

9. Ortiz, *Designing Miracles*, 21. Another example: Penn & Teller titled their television show *Fool Us*.

10. Ortiz, *Designing Miracles*, 30.

11. J. I. Swiss, *Shattering Illusions: Essays on the Ethics, History, and Presentation of Magic* (Seattle: Hermetic Press, 2002), 3–12.

12. J. M. Tyler, R. S. Feldman, and A. Reichert, "The Price of Deceptive Behavior: Disliking and Lying to People Who Lie to Us," *Journal of Experimental Social Psychology* 42, no. 1 (January 2006): 69–77.

13. *Observer*, January 11, 2004, cited in A. Vrij, *Detecting Lies and Deceit: Pitfalls and Opportunities*, 2nd ed. (Chichester, UK: John Wiley and Sons, 2010).

14. S. S. Feldman and E. Cauffman, "Sexual Betrayal among Late Adolescents: Perspectives of the Perpetrator and the Aggrieved," *Journal of Youth and Adolescence* 28, no. 2 (April 1999): 235–258.

15. D. A. Kashy and B. M. DePaulo, "Who Lies?," *Journal of Personality and Social Psychology* 70, no. 5 (May 1996): 1037.

16. Vrij, *Detecting Lies and Deceit*, 32.

17. R. N. Shepard, *Mind Sights: Original Visual Illusions, Ambiguities, and Other Anomalies, with a Commentary on the Play of Mind in Perception and Art* (W. H. Freeman, 1990), 48.

18. In case you don't believe this, you can test this by tracing one table's outline on a thin piece of paper and laying it over the other table.

19. I. Kant, *Kant: Anthropology from a Pragmatic Point of View*, ed. R. B. Louden (Cambridge: Cambridge University Press, 2006), 41.

20. T. Fraps, "Illusions, Magic and the Aesthetics of the Impossible" (unpublished manuscript).

21. M. Spolomon, *Disappearing Tricks: Silent Film, Houdini and the New Magic of the Twentieth Century* (Urbana: University of Illinois Press, 2010), 1.

22. Rensink and Kuhn, "Framework for Using Magic."

23. D. DeEntremont, "Amazing Water Trick! How to Suspend Water Without a Cup!," YouTube video, 2:32, February 13, 2011, https://www.youtube.com/watch?v=7ctaA2mERzI.

24. See B. Tognazzini, "Principles, Techniques, and Ethics of Stage Magic and Their Application to Human Interface Design," in *Proceedings of the INTERACT '93 and CHI '93 Conference on Human Factors in Computing Systems* (New York: ACM, 1993), 355–62.

25. S. Blackmore, *Consciousness: An Introduction* (London: Routledge, 2013).

26. Oxford Living Dictionaries, s.v. "magic," https://en.oxforddictionaries.com/definition/magic.

27. M. Christopher, *The Illustrated History of Magic* (New York: Carroll and Graf, 2006), 12.

28. D. F. Marks, *The Psychology of the Psychic*, 2nd ed. (Amherst, NY: Prometheus Books, 2000).

29. R. Targ and H. Puthoff, "Information Transmission under Conditions of Sensory Shielding," *Nature* 251, no. 5476 (October 18, 1974): 602–7.

30. Marks, *The Psychology of the Psychic*, 153–166.

31. Ranger, J. "Houdini: A Magician among the Spirits," *The Journal of American History* 87, no. 3 (2000): 967–969. doi:10.2307/2675284.

32. See J. Randi, *The Truth about Uri Geller* (Amherst, NY: Prometheus Books, 1982); and Richard Dawkins Foundation for Reason & Science, "Derren Brown Interview (1/6) – Richard Dawkins," YouTube video, 9:52, December 10, 2008, https://www.youtube.com/watch?v=Xswt8B8-UTM.

33. W. Kalush and L. Sloman, *The Secret Life of Houdini: The Making of America's First Superhero* (New York: Atria Books, 2006), 87.

34. D. Brown, "Derren Brown Tricks Advertisers with Subliminal Messaging," YouTube video, 7:02, July 27, 2016, https://www.youtube.com/watch?v=43Mw -f6vIbo.

35. R. Young, "Ep. 72 – Derren Brown," September 9, 2016, in *The Magician's Podcast* produced by R. Young, 2:10:40, http://www.richardyoungmagic.com/ ep-72-derren-brown/.

36. D. Ortiz, *Strong Magic* (n.p.: Ortiz Publications, 1994).

37. See Ortiz, *Strong Magic*, and P. Lamont, *Extraordinary Beliefs: A Historical Approach to a Psychological Problem* (Cambridge: Cambridge University Press, 2013).

38. J. I. Swiss, "Penn and Teller Exposed: An Exclusive Interview," *Genii* 58, no. 7 (May 1995): 491–507.

39. J. Leddington, "The Experience of Magic," *Journal of Aesthetics and Art Criticism* 74, no. 3 (Summer 2016): 253–64.

40. Ortiz, *Designing Miracles*, 30.

41. Leddington, "Experience of Magic," 257.

42. Ortiz, *Strong Magic*, 26.

43. T. S. Gendler, "Alief and Belief," *Journal of Philosophy* 105, no. 10 (October 2008): 635.

44. A. Lane, "Wonderful World: What Walt Disney Made," A Critic at Large, *New Yorker*, December 11, 2006, https://www.newyorker.com/magazine/2006/12/11/ wonderful-world.

45. A. Shtulman and C. Morgan, "The Explanatory Structure of Unexplainable Events: Causal Constraints on Magical Reasoning," *Psychonomic Bulletin & Review* 24, no. 5 (October 2017): 1573–85.

46. B. A. Parris, G. Kuhn, G. A. Mizon, A. Benattayallah, and T. L. Hodgson, "Imaging the Impossible: An fMRI Study of Impossible Causal Relationships in Magic Tricks," *NeuroImage* 45, no. 3 (2009): 1033–1039.

47. For example, see A. H. Danek, M. Öllinger, T. Fraps, B. Grothe, and V. L. Flanagin, "An fMRI Investigation of Expectation Violation in Magic Tricks," *Frontiers in Psychology*, February 4, 2015.

48. Rensink and Kuhn, "Framework for Using Magic."

49. J. D. Woolley, "Thinking about Fantasy: Are Children Fundamentally Different Thinkers and Believers from Adults?," *Child Development* 68, no. 6 (December 1997): 991–1011; and E. Subbotsky, *Magic and the Mind: Mechanisms,*

Functions, and Development of Magical Thinking and Behavior (Oxford: Oxford University Press, 2010), 18–23.

50. J. A. Olson, I. Demacheva, and A. Raz, "Explanations of a Magic Trick across the Life Span," *Frontiers in Psychology*, March 6, 2015.

51. J. Piaget, *The Origins of Intelligence in Children*, trans. M. Cook (New York: International Universities Press, 1952).

52. See G. Kuhn, A. A. Amlani, and R. A. Rensink, "Toward a Science of Magic," *Trends in Cognitive Sciences* 12, no. 9 (September 2008): 349–54.

53. R. Baillargeon and J. DeVos, "Object Permanence in Young Infants: Further Evidence," *Child Development* 62, no. 6 (December 1991): 1227–46.

54. R. B. McCall and P. E. McGhee, "The Discrepancy Hypothesis of Attention and Affect in Infants," in *The Structure of Experience*, ed. I. C. Užgiris and F. Weizmann (New York: Plenum, 1977), 179–210.

55. W. G. Parrott and H. Gleitman, "Infants' Expectations in Play: The Joy of Peek-a-boo," *Cognition and Emotion* 3, no. 4 (1989): 291–311.

56. G. Kuhn and R. Teszka, "Don't Get Misdirected! Differences in Overt and Covert Attentional Inhibition between Children and Adults," *Quarterly Journal of Experimental Psychology* 71, no. 3 (March 2018): 688–94.

57. J. Ahonen, "Taikuutta koirille – Magic for dogs," YouTube video, 1:48, March 21, 2014, https://www.youtube.com/watch?v=VEQXeLjY9ak.

CHAPTER 2: HOW TO CREATE MAGIC

1. J. E. Robert-Houdin, *Memoirs of Robert-Houdin: Ambassador, Author, and Conjuror* (London: Chapman and Hall, 1859).

2. C. Fechner, *The Magic of Robert-Houdin: An Artist's Life*, vol. 1, ed. T. Karr, trans. S. Dagron (Boulogne, France: Éditions F. C. F., 2002).

3. D. Ortiz, *Designing Miracles* (Oakland, CA: Magic Limited, 2006), 37.

4. Ortiz, *Designing Miracles*, 37.

5. J. E. Robert-Houdin, *The Secrets of Conjuring and Magic*, trans. and ed. Professor Hoffman (London: George Routledge and Sons, 1878), 34.

6. P. Lamont, *Extraordinary Beliefs: A Historical Approach to a Psychological Problem* (Cambridge: Cambridge University Press, 2013), 39.

7. See D. Brown, *Confessions of a Conjuror* (London: Channel 4 Books, 2010).

8. J. Grinder and R. Bandler, *The Structure of Magic: A Book about Language and Therapy* (Palo Alto, CA: Science and Behavior Books, 1976).

9. See M. Heap, "The Validity of Some Early Claims of Neuro-linguistic Programming," *Skeptical Intelligencer* 11 (2008): 6–13; and R. Wiseman, C. Watt, L. ten Brinke, S. Porter, S. Couper, and C. Rankin, "The Eyes Don't Have It: Lie Detection and Neuro-linguistic Programming," *PLoS ONE* 7, no. 7 (2012): e40259.

10. "Discipline: 5 Minute 'Speed' Cards," World Memory Statistics, accessed June 11, 2018, http://www.world-memory-statistics.com/discipline.php?id=SPDCARDS.

11. Lamont, *Extraordinary Beliefs*, 42.

12. J. Leddington, "The Experience of Magic," *Journal of Aesthetics and Art Criticism* 74, no. 3 (Summer 2016): 253–64.

13. H. H. Kelley, "Magic Tricks: The Management of Causal Attributions," in *Perspectives on Attribution Research and Theory: The Bielefeld Symposium*, ed. D. Görlitz (Cambridge, MA: Ballinger, 1980), 21.

14. D. Hume, *A Treatise of Human Nature: Being an Attempt to Introduce the Experimental Method of Reasoning into Moral Subjects*, ed. L. A. Selby-Bigge (Oxford: Clarendon Press, 1896), 139; accessed via the University of Michigan Digital Library: https://quod.lib.umich.edu/e/ecco/004806339.0001.001?rgn=main;view=fulltext.

15. A. Michotte, *The Perception of Causality* (New York: Basic Books, 1963).

16. B. J. Scholl and P. D. Tremoulet, "Perceptual Causality and Animacy," *Trends in Cognitive Sciences* 4, no. 8 (August 2000): 299–309.

17. A. M. Leslie and S. Keeble, "Do Six-Month-Old Infants Perceive Causality?," *Cognition* 25, no. 3 (April 1987): 265–88.

18. F. Heider and M. Simmel, "An Experimental Study of Apparent Behavior," *American Journal of Psychology* 57, no. 2 (April 1944): 243–59. For the animation used in their study (which I suggest watching), see "Heider and Simmel Movie," YouTube video, 1:24, posted by "TheIronMagus," January 28, 2009, https://www.youtube.com/watch?v=76p64j3H1Ng.

19. P. Rochat, R. Morgan, and M. Carpenter, "Young Infants' Sensitivity to Movement Information Specifying Social Causality," *Cognitive Development* 12, no. 4 (October–December 1997): 537–61.

20. P. Bertelson, "Ventriloquism: A case of crossmodal perceptual grouping," in *Advances in Psychology*, eds. G. Aschersleben, T. Bachmann, and J. Müsseler (Amsterdam: North-Holland, 1999), 347–362.

21. E. Subbotsky, "Causal Explanations of Events by Children and Adults: Can Alternative Causal Modes Coexist in One Mind?," *British Journal of Developmental Psychology* 19, no. 1 (March 2001): 23–45.

22. Robert-Houdin, *Secrets of Conjuring*, 30.

23. See D. Fitzkee, *The Trick Brain* (San Rafael, CA: San Rafael House, 1944), 21–31; and N. Triplett, "The Psychology of Conjuring Deceptions," *American Journal of Psychology* 11, no. 4 (July 1900): 439–510.

24. S. H. Sharpe, *Conjurers' Psychological Secrets* (Calgary: Hades, 1988).

25. For example, see G. Kuhn, A. A. Amlani, and R. A. Rensink, "Toward a Science of Magic," *Trends in Cognitive Sciences* 12, no. 9 (September 2008): 349–54.

26. P. Lamont and R. Wiseman, *Magic in Theory: An Introduction to the Theoretical and Psychological Elements of Conjuring* (Harpenden, UK: University of Hertfordshire Press, 1999).

27. Kuhn, Amlani, and Rensink, "Toward a Science of Magic."

28. J. Hugard, "Misdirection," *Hugard's Magic Monthly* 17 (1960): 7.

29. See Lamont and Wiseman, *Magic in Theory*, 28.

30. Wikipedia, s.v. "Misdirection (magic)," last modified December 30, 2017, 20:26, https://en.wikipedia.org/wiki/Misdirection_(magic).

31. See G. Kuhn and B. W. Tatler, "Magic and Fixation: Now You Don't See It, Now You Do," *Perception* 34, no. 9 (September 2005): 1155–1161.

32. T. Wonder and S. Minch, *The Books of Wonder*, 2 vols. (Seattle: Hermetic Press, 1994), 1:9–34.

33. See G. Kuhn, B. W. Tatler, J. M. Findlay, and G. G. Cole, "Misdirection in Magic: Implications for the Relationship between Eye Gaze and Attention," *Visual Cognition* 16, no. 2–3 (2008): 391–405.

34. See L. Ganson, *The Magic of Slydini* (Bideford, UK: Supreme Magic, 1980).

35. J. Tamariz, *Magic in Mind: Essential Essays for Magicians*, ed. J. Jay (n.p.: Vanishing, 2012).

36. G. Kuhn, H. A. Caffaratti, R. Teszka, and R. A. Rensink, "A Psychologically-Based Taxonomy of Misdirection," *Frontiers in Psychology*, December 9, 2014.

37. J. Tamariz, *The Magic Way* (Madrid: Editorial Frakson Magic Books, 1988).

38. See C. Thomas and A. Didierjean, "Magicians Fix Your Mind: How Unlikely Solutions Block Obvious Ones," *Cognition* 154 (September 2016): 169–73.

39. C. Thomas, A. Didierjean, and G. Kuhn, "It Is Magic! How Impossible Solutions Prevent the Discovery of Obvious Ones?," *Quarterly Journal of Experimental Psychology*, January 1, 2018.

40. See A. S. Luchins, "Mechanization in Problem Solving: The Effect of *Einstellung*," *Psychological Monographs* 54, no. 6 (1942).

41. See A. Ascanio, "Consideraciones sobre la misdirection," *Misdirection* 1 (1964): 4–6; J. Bruno, *Anatomy of Misdirection* (Baltimore: Stoney Brook Press, 1978; D. Fitzkee, *Magic by Misdirection* (Pomeroy, OH: Lee Jacobs, 1987); A. Leech, *Don't Look Now: The Smart Slant on Misdirection* (Chicago: Ireland Magic, 1960); J. Randal, "Misdirection: The Magician's Insurance," *Genii* 40, no. 6 (June 1976): 380–81; Sharpe, *Conjurers' Psychological Secrets*; and Tamariz, *Magic Way*.

42. Lamont and Wiseman, *Magic in Theory*, 31.

43. See G. Kuhn, B. W. Tatler, and G. G. Cole, "You Look Where I Look! Effect of Gaze Cues on Overt and Covert Attention in Misdirection," *Visual Cognition* 17, no. 6–7 (2009): 925–44.

44. J. Tamariz, *The Five Points to Magic* (Seattle: Hermetic Press, 2007), 35.

45. G. Kuhn, R. Teszka, N. Tenaw, and A. Kingstone, "Don't Be Fooled! Attentional Responses to Social Cues in a Face-to-Face and Video Magic Trick Reveals Greater Top-Down Control for Overt than Covert Attention," *Cognition* 146 (January 2016): 136–42.

46. T. Fraps, "Time and Magic—Manipulating Subjective Temporality," in *Subjective Time: The Philosophy, Psychology, and Neuroscience of Temporality*, ed. V. Arstila and D. Lloyd (Cambridge, MA: MIT Press, 2014), 263–85.

47. Lamont and Wiseman, *Magic in Theory*, 48.

48. R. Wiseman and T. Nakano, "Blink and You'll Miss It: The Role of Blinking in the Perception of Magic Tricks," *PeerJ* 4 (2016): e1873.

49. See E. Ben-Simon, I. Podlipsky, H. Okon-Singer, M. Gruberger, D. Cvetkovic, N. Intrator, and T. Hendler, "The Dark Side of the Alpha Rhythm: fMRI Evidence for Induced Alpha Modulation during Complete Darkness," *European Journal of Neuroscience* 37, no. 5 (March 2013): 795–803.

50. Kuhn et al., "Psychologically-Based Taxonomy."

51. See G. Kuhn and L. M. Martinez, "Misdirection—Past, Present, and the Future," *Frontiers in Human Neuroscience*, January 6, 2012.

52. A. Hergovich, K. Gröbl, and C. Carbon, "The Paddle Move Commonly Used in Magic Tricks as a Means for Analyzing the Perceptual Limits of Combined Motion Trajectories," *Perception* 40, no. 3 (March 2011): 358–66.

53. See R. A. Rensink, "Perception and Attention," in *The Oxford Handbook of Cognitive Psychology*, ed. D. Reisberg (Oxford: Oxford University Press, 2013), 97–116.

54. See T. J. Smith, P. Lamont, and J. M. Henderson, "Change Blindness in a Dynamic Scene Due to Endogenous Override of Exogenous Attentional Cues," *Perception* 42, no. 8 (August 2013): 884–86.

55. J. Shaw, *The Memory Illusion: Remembering, Forgetting, and the Science of False Memory* (London: Random House, 2016).

56. Teller, "Teller Reveals His Secrets," *Smithsonian*, March 2012, https://www.smithsonianmag.com/arts-culture/teller-reveals-his-secrets-100744801/.

57. C. Nolan, dir., *The Prestige* (Burbank, CA: Touchstone Pictures, 2006), DVD.

58. *Genii* 64, no. 8 (August 2001), 48–70.

59. Lamont and Wiseman, *Magic in Theory*, 61.

60. See S. Van de Cruys, J. Wagemans, and V. Ekroll, "The Put-and-Fetch Ambiguity: How Magicians Exploit the Principle of Exclusive Allocation of Movements to Intentions," *i-Perception* 6, no. 2 (April 2015): 86–90.

61. Kuhn et al., "Psychologically-Based Taxonomy."

62. S. Dinsell, dir., *Derren Brown: Miracles for Sale* (London: Objective Media Group, 2011), TV broadcast.

63. See W. Smith, F. Dignum, and L. Sonenberg, "The Construction of Impossibility: A Logic-Based Analysis of Conjuring Tricks," *Frontiers in Psychology*, June 14, 2016.

64. K. R. Popper, *The Logic of Scientific Discovery* (London: Hutchinson, 1959).

CHAPTER 3: THE BELIEF IN REAL MAGIC

1. E. E. Lewis, *A Report of the Mysterious Noises Heard in the House of Mr. John D. Fox, in Hydesville, Arcadia, Wayne County, Authenticated by Certificates, and Confirmed by the Statements of the Citizens of That Place and Vicinity* (Rochester, NY: Shepherd and Reed, 1848).

2. R. B. Davenport, *The Death-Blow to Spiritualism: Being the True Story of the Fox Sisters, as Revealed by Authority of Margaret Fox Kane and Catherine Fox Jencken* (New York: G. W. Dillingham, 1888).

3. See J. M. Henslin, "Craps and Magic," *American Journal of Sociology* 73, no. 3 (November 1967): 316–30.

4. M. Lindeman and A. M. Svedholm, "What's in a Term? Paranormal, Superstitious, Magical and Supernatural Beliefs by Any Other Name Would Mean the Same," *Review of General Psychology* 16, no. 3 (September 2012): 241–55.

5. J. Sørensen, *A Cognitive Theory of Magic* (Lanham, MD: AltaMira, 2007), 9–30.

6. See C. Nemeroff and P. Rozin, "The Makings of the Magical Mind: The Nature and Function of Sympathetic Magical Thinking," in *Imagining the Impossible: Magical, Scientific, and Religious Thinking in Children*, ed. K. S. Rosengren, C. N. Johnson, and P. L. Harris (Cambridge: Cambridge University Press, 2000), 1–34.

7. J. Johnson, "77 Percent of American's Believe in Angels [Polling]," *Inquisitr*, December 23, 2011, https://www.inquisitr.com/171741/77-percent-of-americans -believe-in-angels-polling/.

8. J. L. Risen, "Believing What We Do Not Believe: Acquiescence to Superstitious Beliefs and Other Powerful Intuitions," *Psychological Review* 123, no. 2 (March 2016): 182–207.

9. See G. E. Claridge, *Schizotypy: Implications for Illness and Health* (Oxford: Oxford University Press, 1997); and C. C. French and A. Stone, *Anomalistic Psychology: Exploring Paranormal Belief and Experience* (Basingstoke, UK: Palgrave Macmillan, 2013).

10. French and Stone, *Anomalistic Psychology*, 31.

11. D. W. Moore, "Three in Four Americans Believe in Paranormal," *Gallup*, June 16, 2005, http://news.gallup.com/poll/16915/three-four-americans-believe -paranormal.aspx.

12. C. Campbell, "Half-Belief and the Paradox of Ritual Instrumental Activism: A Theory of Modern Superstition," *British Journal of Sociology* 47, no. 1 (March 1996): 151–166.

13. D. Albas and C. Albas, "Modern Magic: The Case of Examinations," *Sociological Quarterly* 30, no. 4 (Winter 1989): 603–13.

14. See J. L. Bleak and C. M. Frederick, "Superstitious Behavior in Sport: Levels of Effectiveness and Determinants of Use in Three Collegiate Sports," *Journal of Sport Behavior* 21, no. 1 (1998): 1.

15. See V. A. Benassi, B. Singer, and C. B. Reynolds, "Occult Belief: Seeing Is Believing," *Journal for the Scientific Study of Religion* 19, no. 4 (December 1980): 337–49.

16. L. Lesaffre, C. Mohr, D. Rochat, G. Kuhn, and A. Abu-Akel, "Magic Performances—When Explained in Psychic Terms by University Students" (unpublished manuscript).

17. C. Mohr, N. Koutrakis, and G. Kuhn, "Priming Psychic and Conjuring Abilities of a Magic Demonstration Influences Event Interpretation and Random Number Generation Biases," *Frontiers in Psychology*, January 21, 2015.

18. Lesaffre et al., "Magic Performances."

19. K. Simmons, C. Mohr, C. Thomas, and G. Kuhn, "Belief in the Paranormal: Using Pre-exposure Warnings and Alternative Explanations to Debunk Magical Event Interpretation" (poster presentation, Science of Magic Association Conference, Goldsmiths, University of London, August 31, 2017).

20. R. Hodgson and S. J. Davey, "The Possibilities of Mal-observation and Lapse of Memory from a Practical Point of View," *Proceedings of the Society for Psychical Research* 4 (1886–87): 381–495.

21. See also T. Besterman, "The Psychology of Testimony in Relation to Paraphysical Phenomena: Report of an Experiment," *Proceedings of the Society for Psychical Research* 40 (1931–32): 363–387.

22. R. Wiseman, E. Greening, and M. Smith, "Belief in the Paranormal and Suggestion in the Seance Room," *British Journal of Psychology* 94, no. 3 (August 2003): 285–297.

23. K. Wilson and C. C. French, "Magic and Memory: Using Conjuring to Explore the Effects of Suggestion, Social Influence, and Paranormal Belief on Eyewitness Testimony for an Ostensibly Paranormal Event," *Frontiers in Psychology*, November 13, 2014.

24. C. Thomas, C. Mohr, and G. Kuhn, "The Role of Prior Beliefs in the Interpretation of Conjuring Tricks Compared to a Psychic Demonstration" (presentation, Science of Magic Association Conference, Goldsmiths, University of London, August 31–September 1, 2017).

25. Nemeroff and Rozin, "Makings of the Magical Mind."

26. J. G. Frazer, *The Golden Bough: A Study in Magic and Religion* (New York: Macmillan, 1922); and M. Mauss, *A General Theory of Magic*, trans. R. Brain (New York: W. W. Norton, 1972).

27. D. Ullman, "The FDA and Regulation of Homeopathic Medicines," *Huffington Post*, September 23, 2015, https://www.huffingtonpost.com/dana-ullman/the-fda-and-regulation-of_b_8125722.html.

28. British Homeopathic Association, accessed February 7, 2018, https://www.britishhomeopathic.org.

29. Science and Technology Committee, Evidence Check 2: Homeopathy, 2009–10, HC 45 (UK), https://publications.parliament.uk/pa/cm200910/cmselect/cmsctech/45/45.pdf.

30. See A. Shang, K. Huwiler-Müntener, L. Nartey, P. Jüni, S. Dörig, J. A. C. Sterne, D. Pewsner, et al., "Are the Clinical Effects of Homeopathy Placebo Effects? Comparative Study of Placebo-Controlled Trials of Homeopathy and Allopathy," *Lancet* 366, no. 9487 (August 27, 2005): 726–732.

31. Shang et al., "Clinical Effects of Homeopathy."

32. B. M. Hood and P. Bloom, "Children Prefer Certain Individuals over Perfect Duplicates," *Cognition* 106, no. 1 (January 2008): 455–462.

33. M. Park, "Someone Paid $394,000 for J. K. Rowling's Chair," *CNN*, April 7, 2016, https://edition.cnn.com/2016/04/07/entertainment/harry-potter-chair/index.html.

34. R. Wheeler and R. Pocklington, "After Nikki Grahame's Shoe, More Bizarre Celebrity Items on eBay from Niall Horan's Half-Eaten Toast to Britney's Gum," *Mirror*, July 22, 2015, https://www.mirror.co.uk/3am/celebrity-news/one-directions-toast-weirdest-celebrity-6117115.

35. P. Rozin, L. Millman, and C. Nemeroff, "Operation of the Laws of Sympathetic Magic in Disgust and Other Domains," *Journal of Personality and Social Psychology* 50, no. 4 (April 1986): 703–712.

36. Rozin, Millman, and Nemeroff, "Operation of the Laws."

37. P. Rozin, M. Markwith, and B. Ross, "The Sympathetic Magical Law of Similarity, Nominal Realism and Neglect of Negatives in Response to Negative Labels," *Psychological Science* 1, no. 6 (November 1990): 383–384.

38. Frazer, *Golden Bough*, 496.

39. Rozin, Millman, and Nemeroff, "Operation of the Laws."

40. Rozin, Millman, and Nemeroff, "Operation of the Laws."

41. Rozin, Millman, and Nemeroff, "Operation of the Laws."

42. L. A. King, C. M. Burton, J. A. Hicks, and S. M. Drigotas, "Ghosts, UFOs, and Magic: Positive Affect and the Experiential System," *Journal of Personality and Social Psychology* 92, no. 5 (May 2007): 905–919.

43. B. M. Hood, K. Donnelly, U. Leonards, and P. Bloom, "Implicit Voodoo: Electrodermal Activity Reveals a Susceptibility to Sympathetic Magic," *Journal of Cognition and Culture* 10, no. 3–4 (2010): 391–399.

44. B. M. Hood, K. Donnelly, U. Leonards, and P. Bloom, "Implicit Voodoo: Electrodermal Activity Reveals a Susceptibility to Sympathetic Magic," *Journal of Cognition and Culture* 10, no. 3–4 (2010): 391–399.

45. J. Piaget, *The Child's Conception of the World* (New York: Harcourt, Brace, 1929).

46. Piaget, *The Child's Conception of the World*, 135

47. See K. S. Rosengren and A. K. Hickling, "Seeing Is Believing: Children's Explanations of Commonplace, Magical, and Extraordinary Transformations," *Child Development* 65, no. 6 (December 1994): 1605–26.

48. See J. D. Woolley, "Thinking about Fantasy: Are Children Fundamentally Different Thinkers and Believers from Adults?," *Child Development* 68, no. 6 (December 1997): 991–1011.

49. J. D. Woolley, E. A. Boerger, and A. B. Markman, "A Visit from the Candy Witch: Factors Influencing Young Children's Belief in a Novel Fantastical Being," *Developmental Science* 7, no. 4 (September 2004): 456–68.

50. French and Stone, *Anomalistic Psychology*, 77.

51. E. Subbotsky, "Preschool Children's Reception of Unusual Phenomena," *Soviet Psychology* 23, no. 3 (1985): 91–114.

52. E. Subbotsky, "Magical Thinking in Judgments of Causation: Can Anomalous Phenomena Affect Ontological Causal Beliefs in Children and Adults?," *British Journal of Developmental Psychology* 22, no. 1 (March 2004): 123–52.

53. E. Subbotsky, "Causal Explanations of Events by Children and Adults: Can Alternative Causal Modes Coexist in One Mind?," *British Journal of Developmental Psychology* 19, no. 1 (March 2001): 23–45.

54. E. Subbotsky, *Magic and the Mind: Mechanisms, Functions, and Development of Magical Thinking and Behavior* (Oxford: Oxford University Press, 2010, 51).

55. Subbotsky, *Magic and the Mind*, 165.

56. B. Bettelheim, *The Uses of Enchantment: The Meaning and Importance of Fairy Tales* (New York: Vintage Books, 1976).

57. Subbotsky, *Magic and the Mind*, 136.

58. B. Tognazzini, "Principles, Techniques, and Ethics of Stage Magic and Their Application to Human Interface Design," in *Proceedings of the INTERACT '93 and CHI '93 Conference on Human Factors in Computing Systems* (New York: ACM, 1993), 355–362.

59. B. Malinowski, *Magic, Science and Religion and Other Essays* (Boston: Beacon Press, 1948), 30.

60. G. Keinan, "The Effects of Stress and Desire for Control on Superstitious Behavior," *Personality and Social Psychology Bulletin* 28, no. 1 (January 2002): 102–108.

61. J. M. Rudski and A. Edwards, "Malinowski Goes to College: Factors Influencing Students' Use of Ritual and Superstition," *Journal of General Psychology* 134, no. 4 (2007): 389–403.

62. Risen, "Believing What We Do Not."

63. D. Kahneman, *Thinking, Fast and Slow* (New York: Farrar, Strauss and Giroux, 2001).

64. D. Gallagher, "Forget Sharks. These Other Animals Are More Likely to Kill You," *CNET*, June 20, 2015, https://www.cnet.com/news/afraid-of-sharks-these -are-the-animals-more-likely-to-kill-you/.

65. Kahneman, *Thinking, Fast and Slow*.

66. See S. Frederick, "Cognitive Reflection and Decision Making," *Journal of Economic Perspectives* 19, no. 4 (Fall 2005): 25–42.

67. Kahneman, *Thinking, Fast and Slow*; and Frederick, "Cognitive Reflection."

68. Risen, "Believing What We Do Not."

69. Risen, "Believing What We Do Not."

70. R. Davies, "British Gamblers Lost a Record £12.6bn Last Year," *Guardian*, July 1, 2016, https://www.theguardian.com/society/2016/jul/01/british-gamblers -record-losses-fixed-odds-betting.

71. B. Pempus, "Americans Lost $116.9B Gambling in 2016: Report," *Card Player*, February 14, 2017, https://www.cardplayer.com/poker-news/21342 -americans-lost-116-9b-gambling-in-2016-report.

72. Risen, "Believing What We Do Not."

73. V. Denes-Raj and S. Epstein, "Conflict between Intuitive and Rational Processing: When People Behave against Their Better Judgment," *Journal of Personality and Social Psychology* 66, no. 5 (1994): 819.

74. J. P. Simmons and L. D. Nelson, "Intuitive Confidence: Choosing between Intuitive and Nonintuitive Alternatives," *Journal of Experimental Psychology: General* 135, no. 3 (August 2006): 409.

75. See L. Block and T. Kramer, "The Effect of Superstitious Beliefs on Performance Expectations," *Journal of the Academy of Marketing Science* 37, no. 2 (June 2009): 161–69.

76. Risen, "Believing What We Do Not."

CHAPTER 4: THE GAPS IN OUR CONSCIOUS EXPERIENCE

1. Evidence for this comes from drawings on the wall of an Egyptian burial chamber in Beni Hasan, which depict two men kneeling over four inverted bowls. (It is also possible that the men are simply baking bread.) See M. Christopher, *The Illustrated History of Magic* (New York: Carroll and Graf, 2006), 8–29.

2. S. Blackmore, *Consciousness: An Introduction* (London: Routledge, 2013), 55.

3. D. C. Dennett, *Consciousness Explained* (London: Penguin, 1993).

4. Blackmore, *Consciousness*, 61.

5. W. James, *The Principles of Psychology* (1890; repr., New York: Dover, 1950), 1:104.

6. M. F. Land and R. D. Fernald, "The Evolution of Eyes," *Annual Review of Neuroscience* 15 (1992): 1–29.

7. M. F. Land and B. W. Tatler, *Looking and Acting: Vision and Eye Movements in Natural Behaviour* (Oxford: Oxford University Press, 2009).

8. M. F. Land, J. N. Marshall, D. Brownless, and T. W. Cronin, "The Eye-Movements of the Mantis Shrimp *Odontodactylus scyllarus* (Crustacea: Stomatopoda)," *Journal of Comparative Physiology A* 167, no. 2 (July 1990): 155–66.

9. M. Land, N. Mennie, and J. Rusted, "The Roles of Vision and Eye Movements in the Control of Activities of Daily Living," *Perception* 28, no. 11 (November 1999): 1311–28.

10. G. Kuhn and B. W. Tatler, "Magic and Fixation: Now You Don't See It, Now You Do," *Perception* 34, no. 9 (September 2005): 1155–61.

11. R. Wiseman and T. Nakano, "Blink and You'll Miss It: The Role of Blinking in the Perception of Magic Tricks," *PeerJ* 4 (2016): e1873.

12. Kuhn and Tatler, "Magic and Fixation"; G. Kuhn, B. W. Tatler, and G. G. Cole, "You Look Where I Look! Effect of Gaze Cues on Overt and Covert Attention in Misdirection," *Visual Cognition* 17, no. 6–7 (2009): 925–44; and G. Kuhn, B. W. Tatler, J. M. Findlay, and G. G. Cole, "Misdirection in Magic: Implications for the Relationship between Eye Gaze and Attention," *Visual Cognition* 16, no. 2–3 (2008): 391–405.

13. D. Melcher and C. L. Colby, "Trans-saccadic Perception," *Trends in Cognitive Sciences* 12, no. 12 (December 2008): 466–73.

14. Kuhn and Tatler, "Magic and Fixation"; Kuhn, Tatler, and Cole, "You Look Where I Look"; Kuhn et al., "Misdirection in Magic"; and G. Kuhn and B. W. Tatler, "Misdirected by the Gap: The Relationship between Inattentional Blindness and Attentional Misdirection," *Consciousness and Cognition* 20, no. 2 (June 2011): 432–36.

15. A. S. Barnhart and S. D. Goldinger, "Blinded by Magic: Eye-Movements Reveal the Misdirection of Attention," *Frontiers in Psychology*, December 17, 2014.

16. P. Lamont and R. Wiseman, *Magic in Theory: An Introduction to the Theoretical and Psychological Elements of Conjuring* (Harpenden, UK: University of Hertfordshire Press, 1999).

17. S. P. Liversedge and J. M. Findlay, "Saccadic Eye Movements and Cognition," *Trends in Cognitive Sciences* 4, no. 1 (January 2000): 6–14.

18. K. Fonseca-Azevedo and S. Herculano-Houzel, "Metabolic Constraint Imposes Tradeoff between Body Size and Number of Brain Neurons in Human Evolution," *Proceedings of the National Academy of Sciences of the United States of America* 109, no. 45 (November 6, 2012): 18571–76.

19. F. Du, X. Zhu, Y. Zhang, M. Friedman, N. Zhang, K. Uğurbil, and W. Chen, "Tightly Coupled Brain Activity and Cerebral ATP Metabolic Rate," *Proceedings of the National Academy of Sciences of the United States of America* 105, no. 17 (April 29, 2008): 6409–14.

20. A. L. Yarbus, *Eye Movements and Vision* (New York: Plenum, 1967), 171–196.

21. See M. F. Land, "Eye Movements and the Control of Actions in Everyday Life," *Progress in Retinal and Eye Research* 25, no. 3 (May 2006): 296–324.

22. R. A. Rensink, "The Dynamic Representation of Scenes," *Visual Cognition* 7, no. 1–3 (2000): 17–42.

23. N. J. Emery, "The Eyes Have It: The Neuroethology, Function and Evolution of Social Gaze," *Neuroscience & Biobehavioral Reviews* 24, no. 6 (August 2000): 581–604.

24. See T. Farroni, G. Csibra, G. Simion, and M. H. Johnson, "Eye Contact Detection in Humans from Birth," *Proceedings of the National Academy of Sciences of the United States of America* 99, no. 14 (July 9, 2002): 9602–5; and V. Corkum and C. Moore, "The Origins of Joint Visual Attention in Infants," *Developmental Psychology* 34, no. 1 (January 1998): 28–38.

25. S. Fletcher-Watson, S. R. Leekam, V. Benson, M. C. Frank, and J. M. Findlay, "Eye-Movements Reveal Attention to Social Information in Autism Spectrum Disorder," *Neuropsychologia* 47, no. 1 (2009): 248–57.

26. G. Kuhn and A. Kingstone, "Look Away! Eyes and Arrows Engage Oculomotor Responses Automatically," *Attention, Perception, & Psychophysics* 71, no. 2 (February 2009): 314–27.

27. B. W. Tatler and G. Kuhn, "Don't Look Now: The Magic of Misdirection," in *Eye Movements: A Window on Mind and Brain*, ed. R. P. G. van Gompel, M. H. Fischer, W. S. Murray, and R. L. Hill (Oxford: Elsevier, 2007), 697–714.

28. Kuhn, Tatler, and Cole, "You Look Where I Look."

29. A. Hergovich and B. Oberfichtner, "Magic and Misdirection: The Influence of Social Cues on the Allocation of Visual Attention while Watching a Cups-and-Balls Routine," *Frontiers in Psychology*, May 31, 2016.

30. G. Kuhn, R. Teszka, N. Tenaw, and A. Kingstone, "Don't Be Fooled! Attentional Responses to Social Cues in a Face-to-Face and Video Magic Trick Reveals Greater Top-Down Control for Overt than Covert Attention," *Cognition* 146 (January 2016): 136–42.

31. H. Scott, J. P. Batton, and G. Kuhn, "Why Are You Looking at Me? It's Because I'm Talking, but Mostly Because I'm Staring or Not Doing Much" (unpublished manuscript).

32. J. M. Findlay and I. D. Gilchrist, *Active Vision: The Psychology of Looking and Seeing* (Oxford: Oxford University Press, 2003).

33. Lamont and Wiseman, *Magic in Theory*, 37.

34. Barnhart and S. D. Goldinger, "Blinded by Magic."

35. T. J. Smith, P. Lamont, and J. M. Henderson, "The Penny Drops: Change Blindness at Fixation," *Perception* 41, no. 4 (April 2012): 489–92.

36. Kuhn et al., "Don't Be Fooled."

37. E. C. Cherry, "Some Experiments on the Recognition of Speech with One and Two Ears," *Journal of the Acoustical Society America* 25, no. 5 (September 1953): 975–79.

38. D. J. Simons and C. F. Chabris, "Gorillas in Our Midst: Sustained Inattentional Blindness for Dynamic Events," *Perception* 28, no. 9 (September 1999): 1059–74.

39. U. Neisser, "The Control of Information Pickup in Selective Looking," in *Perception and Its Development: A Tribute to Eleanor J. Gibson*, ed. E. J. Gibson and A. D. Pick (Hillsdale, NJ: Lawrence Erlbaum, 1979), 201–19.

40. A. Mack and I. Rock, *Inattentional Blindness* (Cambridge, MA: MIT Press, 1998).

41. S. B. Most, "What's 'Inattentional' about Inattentional Blindness?," *Consciousness and Cognition* 19, no. 4 (December 2010): 1102–4; and M. S. Jensen, R. Yao, W. N. Street, and D. J. Simons, "Change Blindness and Inattentional Blindness," *WIREs Cognitive Science* 2, no. 5 (September/October 2011): 529–46.

42. D. Memmert, D. J. Simons, and T. Grimme, "The Relationship between Visual Attention and Expertise in Sports," *Psychology of Sport and Exercise* 10, no. 1 (January 2009): 146–51.

43. G. W. McConkie and D. Zola, "Is Visual Information Integrated across Successive Fixations in Reading?," *Perception & Psychophysics* 25, no. 3 (May 1979): 221–24.

44. J. Grimes, "On the Failure to Detect Changes in Scenes across Saccades," in *Perception*, ed. K. Akins, Vancouver Studies in Cognitive Science 5 (Oxford: Oxford University Press, 1996), 89–110.

45. R. A. Rensink, J. K. O'Regan, and J. J. Clark, "To See or Not to See: The Need for Attention to Perceive Changes in Scenes," *Psychological Science* 8, no. 5 (September 1997): 368–73.

46. R. A. Rensink, "Change Detection," *Annual Review of Psychology* 53 (2002): 245–77.

47. D. T. Levin, D. J. Simons, B. L. Angelone, and C. F. Chabris, "Memory for Centrally Attended Changing Objects in an Incidental Real-World Change Detection Paradigm," *British Journal of Psychology* 93, no. 3 (August 2002): 289–302.

48. D. T. Levin and D. J. Simons, "Failure to Detect Changes to Attended Objects in Motion Pictures," *Psychonomic Bulletin & Review* 4, no. 4 (December 1997): 501–6.

49. D. T. Levin and M. R. Beck, "Thinking about Seeing: Spanning the Difference between Metacognitive Failure and Success," in *Thinking and Seeing: Visual Metacognition in Adults and Children*, ed. D. T. Levin (Cambridge, MA: MIT Press, 2004), 121–43.

50. Kuhn and Tatler, "Misdirected by the Gap."

51. D. Memmert, "The Gap between Inattentional Blindness and Attentional Misdirection," *Consciousness and Cognition* 19, no. 4 (December 2010): 1097–101; D. Memmert and P. Furley, "Beyond Inattentional Blindness and Attentional Misdirection: From Attentional Paradigms to Attentional Mechanisms," *Consciousness and Cognition* 19, no. 4 (December 2010): 1107–9; and Kuhn and Tatler, "Misdirected by the Gap"; and Most, "What's 'Inattentional.'"

52. Kuhn and Tatler, "Magic and Fixation"; and Kuhn et al., "Misdirection in Magic."

53. Kuhn et al., "Misdirection in Magic."

54. Kuhn et al., "Misdirection in Magic."

55. E. Fischer, R. F. Haines, and T. A. Price, *Cognitive Issues in Head-Up Displays*, NASA Technical Paper 1711 (n.p.: National Aeronautics and Space Administration, Scientific and Technical Information Branch, 1980).

56. I. E. Hyman Jr., S. M. Boss, B. M. Wise, K. E. McKenzie, and J. M. Caggiano, "Did You See the Unicycling Clown? Inattentional Blindness while Walking and Talking on a Cell Phone," *Applied Cognitive Psychology* 24, no. 5 (July 2010): 597–607.

57. D. L. Strayer, F. A. Drews, and D. J. Crouch, "A Comparison of the Cell Phone Driver and the Drunk Driver," *Human Factors* 48, no. 2 (Summer 2006): 381–91.

58. D. Crundall, M. Bains, P. Chapman, and G. Underwood, "Regulating Conversation during Driving: A Problem for Mobile Telephones?," *Transportation Research Part F: Traffic Psychology and Behavior* 8, no. 3 (May 2005): 197–211; and D. L. Strayer and W. A. Johnston, "Driven to Distraction: Dual-Task Studies of Simulated Driving and Conversing on a Cellular Telephone," *Psychological Science* 12, no. 6 (November 2001): 462–66.

59. Strayer, Drews, and Crouch, "Comparison of the Cell Phone Driver."

60. Strayer, Drews, and Crouch, "Comparison of the Cell Phone Driver."

61. C. Stothart, A. Mitchum, and C. Yehnert, "The Attentional Cost of Receiving a Cell Phone Notification," *Journal of Experimental Psychology: Human Perception and Performance* 41, no. 4 (August 2015): 893.

62. B. J. Dixon, M. J. Daly, H. Chan, A. D. Vescan, I. J. Witterick, and J. C. Irish, "Surgeons Blinded by Enhanced Navigation: The Effect of Augmented Reality on Attention," *Surgical Endoscopy* 27, no. 2 (February 2013): 454–61.

63. C. B. Jaeger, D. T. Levin, and E. Porter, "Justice Is (Change) Blind: Applying Research on Visual Metacognition in Legal Settings," *Psychology, Public Policy, and Law* 23, no. 2 (May 2017): 259–79.

64. C. F. Chabris and D. J. Simons, *The Invisible Gorilla: How Our Intuitions Deceive Us* (New York: Crown, 2009), 1–11.

65. C. F. Chabris, A. Weinberger, M. Fontaine, and D. J. Simons, "You Do Not Talk about Fight Club if You Do Not Notice Fight Club: Inattentional Blindness for a Simulated Real-World Assault," *i-Perception* 2, no. 2 (February 2011): 150–53.

CHAPTER 5: SEEING IS BELIEVING

1. H. von Helmholtz, *Treatise on Physiological Optics*, ed. J. P. C. Southall, vol. 3, *The Perception of Vision* (1924; repr., Mineola, NY: Dover, 2005).

2. G. Mather, *The Psychology of Visual Art: Eye, Brain and Art* (Cambridge: Cambridge University Press, 2014), 6.

3. G. Lamb, *Victorian Magic* (London: Routledge and Kegan Paul, 1976), 41–50.

4. R. L. Gregory, *Seeing through Illusions* (Oxford: Oxford University Press, 2009), 127.

5. J. Spencer, J. O'Brien, P. Heard, and R. Gregory, "Do Infants See the Hollow Face Illusion?" (poster presentation, 34th European Conference on Visual Perception, Toulouse, France, September 1, 2011).

6. D. Dima, J. P. Roiser, D. E. Dietrich, C. Bonnemann, H. Lanfermann, H. M. Emrich, and W. Dillo, "Understanding Why Patients with Schizophrenia Do Not Perceive the Hollow-Mask Illusion Using Dynamic Causal Modeling," *NeuroImage* 46, no. 4 (July 15, 2009): 1180–86.

7. E. Pellicano and D. Burr, "When the World Becomes 'Too Real': A Bayesian Explanation of Autistic Perception," *Trends in Cognitive Sciences* 16, no. 10 (October 2012): 504–10.

8. See G. Kuhn, A. Kourkoulou, and S. R. Leekam, "How Magic Changes Our Expectations about Autism," *Psychological Science* 21, no. 10 (October 2010): 1487–93.

9. Gregory, *Seeing through Illusions*.

10. V. Ekroll, B. Sayim, and J. Wagemans, "The Other Side of Magic: The Psychology of Perceiving Hidden Things," *Perspectives on Psychological Science* 12, no. 1 (January 2017): 91–106.

11. A. S. Barnhart, "The Exploitation of Gestalt Principles by Magicians," *Perception* 39, no. 9 (September 2010): 1286–89.

12. Ekroll, Sayim, and Wagemans, "Other Side of Magic."

13. V. Ekroll, B. Sayim, and J. Wagemans, "Against Better Knowledge: The Magical Force of Amodal Volume Completion," *i-Perception* 4, no. 8 (December 2013): 511–15.

14. V. Ekroll, B. Sayim, R. Van der Hallen, and J. Wagemans, "Illusory Visual Completion of an Object's Invisible Backside Can Make Your Finger Feel Shorter," *Current Biology* 26, no. 8 (April 25, 2016): 1029–33.

15. Ekroll, Sayim, and Wagemans, "Other Side of Magic."

16. Ekroll, Sayim, and Wagemans, "Other Side of Magic."

17. The illusion was first described by Max Dessoir in 1893. N. Triplett, "The Psychology of Conjuring Deceptions," *American Journal of Psychology* 11, no. 4 (July 1900): 439–510.

18. Triplett, "Psychology of Conjuring Deceptions," 492.

19. G. Kuhn and M. F. Land, "There's More to Magic than Meets the Eye," *Current Biology* 16, no. 22 (November 21, 2006): R950–51.

20. G. Kuhn and R. A. Rensink, "The Vanishing Ball Illusion: A New Perspective on the Perception of Dynamic Events," *Cognition* 148 (March 2016): 64–70.

21. G. Underwood, "Visual Attention and the Transition from Novice to Advanced Driver," *Ergonomics* 50, no. 8 (2007): 1235–49.

22. A. D. Milner and M. A. Goodale, *The Visual Brain in Action* (Oxford: Oxford University Press, 1995).

23. G. Króliczak, P. Heard, M. A. Goodale, and R. L. Gregory, "Dissociation of Perception and Action Unmasked by the Hollow-Face Illusion," *Brain Research* 1080 (March 29, 2006): 9–16.

24. M. Changizi, *The Vision Revolution* (Dallas: BenBella Books, 2009), 109–162.

25. B. Khurana and R. Nijhawan, "Extrapolation or Attention Shift," *Nature* 378, no. 6557 (December 7, 1995): 566.

26. J. J. Freyd and R. A. Finke, "Representational Momentum," *Journal of Experimental Psychology: Learning Memory and Cognition* 10, no. 1 (January 1984): 126–32.

27. Kuhn and Rensink, "Vanishing Ball Illusion."

28. J. Cui, J. Otero-Millan, S. L. Macknik, M. King, and S. Martinez-Conde, "Social Misdirection Fails to Enhance a Magic Illusion," *Frontiers in Human Neuroscience*, September 29, 2011.

29. M. L. Tompkins, A. T. Woods, and A. M. Aimola Davies, "Phantom Vanish Magic Trick: Investigating the Disappearance of a Non-existent Object in a Dynamic Scene," *Frontiers in Psychology*, July 21, 2016.

30. F. G. E. Happé, "Studying Weak Central Coherence at Low Levels: Children with Autism Do Not Succumb to Visual Illusions. A Research Note," *Journal of Child Psychology and Psychiatry* 37, no. 7 (October 1996): 873–77.

31. S. Baron-Cohen, *Mindblindness: An Essay on Autism and Theory of Mind* (Cambridge, MA: MIT Press, 1995).

32. Kuhn, Kourkoulou, and Leekam, "How Magic Changes."

33. M. Ruggeri, J. C. Major Jr., C. McKeown, R. W. Knighton, C. A. Puliafito, and S. Jiao, "Retinal Structure of Birds of Prey Revealed by Ultra-high Resolution Spectral-Domain Optical Coherence Tomography," *Investigative Ophthalmology & Visual Science* 51, no. 11 (November 2010): 5789–95.

CHAPTER 6: MEMORY ILLUSIONS

1. P. Lamont and R. Wiseman, "The Rise and Fall of the Indian Rope Trick," *Journal of the Society for Psychical Research* 65 (2001): 175–93.

2. P. Lamont, *The Rise of the Indian Rope Trick: The Biography of a Legend* (London: Little, Brown, 2004), 153–172.

3. Quoted in Lamont and Wiseman, "Rise and Fall of the Indian Rope Trick," 7.

4. R. Hodgson and S. J. Davey, "The Possibilities of Mal-observation and Lapse of Memory from a Practical Point of View," *Proceedings of the Society for Psychical Research* 4 (1886–87): 381–495.

5. C. F. Chabris and D. J. Simons, *The Invisible Gorilla: How Our Intuitions Deceive Us* (New York: Crown, 2009), 45.

6. E. Tulving and D. M. Thomson, "Encoding Specificity and Retrieval Processes in Episodic Memory," *Psychological Review* 80, no. 5 (September 1973): 352.

7. J. Tamariz, "Fundamentals in Illusionism," in *Magic in Mind: Essential Essays for Magicians*, ed. J. Jay (n.p.: Vanishing, 2012), 161–162.

8. R. A. Rensink, J. K. O'Regan, and J. J. Clark, "To See or Not to See: The Need for Attention to Perceive Changes in Scenes," *Psychological Science* 8, no. 5 (September 1997): 368–73.

9. D. T. Levin, S. B. Drivdahl, N. Momen, and M. R. Beck, "False Predictions about the Detectability of Visual Changes: The Role of Beliefs about Attention, Memory, and the Continuity of Attended Objects in Causing Change Blindness Blindness," *Consciousness and Cognition* 11, no. 4 (December 2002): 507–27.

10. R. S. Nickerson and M. J. Adams, "Long-Term-Memory for a Common Object," *Cognitive Psychology* 11, no. 3 (July 1979): 287–307.

11. R. Brown and J. Kulik, "Flashbulb Memories," *Cognition* 5, no. 1 (1977): 73–99.

12. K. Pezdek, "Event Memory and Autobiographical Memory for the Events of September 11, 2001," *Applied Cognitive Psychology* 17, no. 9 (November/December 2003): 1033–45.

13. J. M. Talarico and D. C. Rubin, "Confidence, Not Consistency, Characterizes Flashbulb Memories," *Psychological Science* 14, no. 5 (September 2003): 455–61.

14. F. C. Bartlett, *Remembering: An Experimental and Social Study* (Cambridge: Cambridge University Press, 1932).

15. H. L. Roediger and K. B. McDermott, "Creating False Memories: Remembering Words Not Presented in Lists," *Journal of Experimental Psychology: Learning Memory and Cognition* 21, no. 4 (July 1995): 803–14.

16. E. F. Loftus and J. C. Palmer, "Reconstruction of Automobile Destruction: An Example of the Interaction between Language and Memory," *Journal of Verbal Learning and Verbal Behavior* 13, no. 5 (October 1974): 585–89.

17. M. Garry, C. G. Manning, E. F. Loftus, and S. J. Sherman, "Imagination Inflation: Imagining a Childhood Event Inflates Confidence that It Occurred," *Psychonomic Bulletin & Review* 3, no. 2 (June 1996): 208–14.

18. K. A. Braun, R. Ellis, and E. F. Loftus, "Make My Memory: How Advertising Can Change Our Memories of the Past," *Psychology & Marketing* 19, no.1 (January 2002): 1–23.

19. E. F. Loftus, "Our Changeable Memories: Legal and Practical Implications," *Nature Reviews Neuroscience* 4, no. 3 (March 2003): 231.

20. K. A. Wade, M. Garry, J. D. Read, and D. S. Lindsay, "A Picture Is Worth a Thousand Lies: Using False Photographs to Create False Childhood Memories," *Psychonomic Bulletin & Review* 9, no. 3 (September 2002): 597–603.

21. R. Wiseman and E. Greening, "'It's Still Bending': Verbal Suggestion and Alleged Psychokinetic Ability," *British Journal of Psychology* 96, no. 1 (February 2005): 115–27.

22. G. Kuhn and M. F. Land, "There's More to Magic than Meets the Eye," *Current Biology* 16, no. 22 (November 21, 2006): R950–51.

23. Garry et al., "Imagination Inflation."

24. A. K. Thomas and E. F. Loftus, "Creating Bizarre False Memories through Imagination," *Memory & Cognition* 30, no. 3 (April 2002): 423–31.

25. S. E. Asch, "Studies of Independence and Conformity: I. A Minority of One against a Unanimous Majority," *Psychological Monographs: General and Applied* 70, no. 9 (1956): 1.

26. D. M. Schneider and M. J. Watkins, "Response Conformity in Recognition Testing," *Psychonomic Bulletin & Review* 3, no. 4 (December 1996): 481–85.

27. F. Gabbert, A. Memon, and K. Allan, "Memory Conformity: Can Eyewitnesses Influence Each Other's Memories for an Event?," *Applied Cognitive Psychology* 17, no. 5 (July 2003): 533–43.

CHAPTER 7: MIND CONTROL AND THE MAGICIAN'S FORCE

1. See H. Poincaré, *Science et Méthode* (Paris: Flammarion, 1908).

2. E. Lorenz, "Predictability: Does the Flap of a Butterfly's Wing in Brazil Set Off a Tornado in Texas?" (lecture, 139th Meeting of the American Association for the Advancement of Science, Washington, DC, December 29, 1972).

3. For example, see J. McFadden and J. Al-Khalili, *Life on the Edge: The Coming of Age of Quantum Biology* (New York: Broadway Books, 2016).

4. See M. S. Gazzaniga, *Who's in Charge? Free Will and the Science of the Brain* (London: Robinson, 2012).

5. For example, see D. M. Wegner, "The Mind's Best Trick: How We Experience Conscious Will," *Trends in Cognitive Sciences* 7, no. 2 (February 2003): 65–69.

6. G. Kuhn, A. A. Amlani, and R. A. Rensink, "Toward a Science of Magic," *Trends in Cognitive Sciences* 12, no. 9 (September 2008): 349–54.

7. J. A. Olson, A. A. Amlani, A. Raz, and R. A. Rensink, "Influencing Choice without Awareness," *Consciousness and Cognition* 37 (December 2015): 225–36.

8. D. E. Shalom, M. G. de Sousa Serro, M. Giaconia, L. M. Martinez, A. Rieznik, and M. Sigman, "Choosing in Freedom or Forced to Choose? Introspective Blindness to Psychological Forcing in Stage-Magic," *PloS ONE* 8, no. 3 (2013): e58254.

9. J. A. Olson, A. A. Amlani, and R. A. Rensink, "Perceptual and Cognitive Characteristics of Common Playing Cards," *Perception* 41, no. 3 (March 2012): 268–86.

10. Shalom et al., "Choosing in Freedom."

11. H. Ozono, "What Kind of Magician's Choice Is More Effective? An Experiment" (poster presentation, Science of Magic Association Conference, Goldsmiths, University of London, August 31, 2017).

12. Shalom et al., "Choosing in Freedom."

13. K. Eschner, "What We Know about the CIA's Midcentury Mind-Control Project," *Smithsonian*, April 13, 2017, https://www.smithsonianmag.com/smart-news/what-we-know-about-cias-midcentury-mind-control-project-180962836/.

14. R. Masters and J. Houston, *The Varieties of Psychedelic Experience: The Classic Guide to the Effects of LSD on the Human Psyche* (Rochester, VT: Park Street Press, 2000).

15. B. Sidis, *The Psychology of Suggestion* (New York: D. Appleton, 1898), 24–44; accessed via https://www.sidis.net/psychologyofsugg.pdf.

16. M. Stroh, A. M. Shaw, and M. F. Washburn, "A Study in Guessing," *American Journal of Psychology* 19, no. 2 (April 1908): 243–45.

17. See J. K. Adams, "Laboratory Studies of Behavior without Awareness," *Psychological Bulletin* 54, no. 5 (September 1957): 383.

18. W. M. O'Barr, "'Subliminal' Advertising," *Advertising & Society Review* 6, no. 4 (2005): fig. 3, https://muse.jhu.edu/article/193862/.

19. O'Barr, "'Subliminal' Advertising."

20. O'Barr, "'Subliminal' Advertising."

21. See C. W. Eriksen, "Discrimination and Learning without Awareness: A Methodological Survey and Evaluation," *Psychological Review* 67, no. 5 (September 1960): 279.

22. Eriksen, "Discrimination and Learning."

23. See J. C. Karremans, W. Stroebe, and J. Claus, "Beyond Vicary's Fantasies: The Impact of Subliminal Priming and Brand Choice," *Journal of Experimental Social Psychology* 42, no. 6 (November 2006): 792–98; A. G. Greenwald, E. R. Spangenberg, A. R. Pratkanis, and J. Eskenazi, "Double-Blind Tests of Subliminal Self-Help Audiotapes," *Psychological Science* 2, no. 2 (March 1991): 119–22; and P. M. Merikle and H. E. Skanes, "Subliminal Self-Help Audiotapes: A Search for Placebo Effects," *Journal of Applied Psychology* 77, no. 5 (October 1992): 772.

24. S. Kouider and S. Dehaene, "Levels of Processing during Non-conscious Perception: A Critical Review of Visual Masking," *Philosophical Transactions of the Royal Society B: Biological Sciences* 362, no. 1481 (May 29, 2007): 857–75.

25. J. S. Morris, A. Öhman, and R. J. Dolan, "Conscious and Unconscious Emotional Learning in the Human Amygdala," *Nature* 393, no. 6684 (June 4, 1998): 467.

26. A. Mack and I. Rock, *Inattentional Blindness* (Cambridge, MA: MIT Press, 1998).

27. A. Berti and G. Rizzolatti, "Visual Processing without Awareness: Evidence from Unilateral Neglect," *Journal of Cognitive Neuroscience* 4, no. 4 (Fall 1992): 345–51.

28. W. B. Key, *Subliminal Seduction: Ad Media's Manipulation of Not So Innocent America* (Englewood Cliffs, NJ: Prentice-Hall, 1973).

29. J. A. Bargh, "What Have We Been Priming All These Years? On the Development, Mechanisms, and Ecology of Nonconscious Social Behavior," *European Journal of Social Psychology* 36, no. 2 (March/April 2006): 147–68.

30. T. K. Srull and R. S. Wyer, "The Role of Category Accessibility in the Interpretation of Information about Persons: Some Determinants and Implications," *Journal of Personality and Social Psychology* 37, no. 10 (October 1979): 1660.

31. J. A. Bargh, M. Chen, and L. Burrows, "Automaticity of Social Behavior: Direct Effects of Trait Construct and Stereotype Activation on Action," *Journal of Personality and Social Psychology* 71, no. 2 (August 1996): 230–44.

32. S. Doyen, O. Klein, C. Pichon, and A. Cleeremans, "Behavioral Priming: It's All in the Mind, but Whose Mind?," *PloS ONE* 7, no. 1 (2012): e29081.

33. C. Trappey, "A Meta-analysis of Consumer Choice and Subliminal Advertising," *Psychology & Marketing* 13, no. 5 (August 1996): 517–30.

34. Wegner, "Mind's Best Trick."

35. M. Faraday, "Experimental Investigation of Table-Moving," *Athenæum, Journal of English Foreign Literature, Science, and the Fine Arts*, no. 1340 (July 2, 1853): 801–3.

36. H. L. Gauchou, R. A. Rensink, and S. Fels, "Expression of Nonconscious Knowledge via Ideomotor Actions," *Consciousness and Cognition* 21, no. 2 (June 2012): 976–82.

37. W. Penfield, *The Mystery of Mind: A Critical Study of Consciousness and the Human Mind* (Princeton, NJ: Princeton University Press, 1975).

38. Wegner, "Mind's Best Trick."

39. B. Libet, "Unconscious Cerebral Initiative and the Role of Conscious Will in Voluntary Action," *Behavioral and Brain Sciences* 8, no. 4 (December 1985): 529–39.

40. C. S. Soon, M. Brass, H. Heinze, and J. Haynes, "Unconscious Determinants of Free Decisions in the Human Brain," *Nature Neuroscience* 11, no. 5 (May 2008): 543–45.

41. J. A. Olson, M. Landry, K. Appourchaux, and A. Raz, "Simulated Thought Insertion: Influencing the Sense of Agency Using Deception and Magic," *Consciousness and Cognition* 43 (July 2016): 11–26.

42. D. M. Wegner and T. Wheatley, "Apparent Mental Causation: Sources of the Experience of Will," *American Psychologist* 54, no. 7 (July 1999): 480–92.

43. Wegner and Wheatley, "Apparent Mental Causation."

44. Shalom et al., "Choosing in Freedom."

45. R. E. Nisbett and T. D. Wilson, "Telling More than We Know: Verbal Reports on Mental Processes," *Psychological Review* 84, no. 3 (May 1977): 231–59.

46. P. Johansson, L. Hall, S. Sikström, and A. Olsson, "Failure to Detect Mismatches between Intention and Outcome in a Simple Decision Task," *Science* 310, no. 5745 (October 7, 2005): 116–19.

47. L. Hall, P. Johansson, B. Tärning, S. Sikström, and T. Deutgen, "Magic at the Marketplace: Choice Blindness for the Taste of Jam and the Smell of Tea," *Cognition* 117, no. 1 (October 2010): 54–61.

48. L. Hall, P. Johansson, and T. Strandberg, "Lifting the Veil of Morality: Choice Blindness and Attitude Reversals on a Self-Transforming Survey," *PLoS ONE* 7, no. 9 (2012): e45457.

49. See A. R. Pratkanis and E. Aronson, *Age of Propaganda: The Everyday Use and Abuse of Persuasion*, rev. ed. (New York: Henry Holt, 2001).

CHAPTER 8: MIND CONTROL THROUGH HYPNOSIS

1. D. Brown, "Derren Brown on the London Underground," YouTube video, 1:33, posted November 28, 2011, https://www.youtube.com/watch?v=Sg436FRLPX0.

2. H. J. Crawford, M. Kitner-Triolo, S. W. Clarke, and B. Olesko, "Transient Positive and Negative Experiences Accompanying Stage Hypnosis," *Journal of Abnormal Psychology* 101, no. 4 (November 1992): 663.

3. C. Musès and A. M. Young, *Consciousness and Reality: The Human Pivot Point* (n.p.: Avon Books, 1983), 14.

4. A. Gauld, *A History of Hypnotism* (Cambridge: Cambridge University Press, 1992), 9.

5. J. F. Kihlstrom, "Mesmer, the Franklin Commission, and Hypnosis: A Counterfactual Essay," *International Journal of Clinical and Experimental Hypnosis* 50, no. 4 (2002): 407–19.

6. E. R. Hilgard, *Divided Consciousness: Multiple Controls in Human Thought and Action* (New York: Wiley, 1977).

7. M. R. Nash and A. J. Barnier, eds., *The Oxford Handbook of Hypnosis: Theory, Research, and Practice* (Oxford: Oxford University Press, 2012).

8. N. P. Spanos, "Hypnotic Behavior: A Social-Psychological Interpretation of Amnesia, Analgesia, and 'Trance Logic,'" *Behavioral and Brain Sciences* 9, no. 3 (September 1986): 449–67.

9. M. T. Orne, "On the Social Psychology of the Psychological Experiment: With Particular Reference to Demand Characteristics and Their Implications," *American Psychologist* 17, no. 11 (November 1962): 776–83.

10. A. J. de Craen, T. J. Kaptchuk, J. G. Tijssen, and J. Kleijnen, "Placebos and Placebo Effects in Medicine: Historical Overview," *Journal of the Royal Society of Medicine* 92, no. 10 (October 1999): 511–15.

11. M. T. Orne and F. J. Evans, "Social Control in Psychological Experiment: Antisocial Behavior and Hypnosis," *Journal of Personality and Social Psychology* 1, no. 3 (March 1965): 189–200.

12. K. S. Bowers, "The Waterloo-Stanford Group C (WSGC) Scale of Hypnotic Susceptibility: Normative and Comparative Data," *International Journal of Clinical and Experimental Hypnosis* 41, no. 1 (1993): 35–46.

13. P. W. Sheehan and D. Statham, "Associations between Lying and Hypnosis: An Empirical Analysis," *British Journal of Experimental & Clinical Hypnosis* 5, no. 2 (1988): 87–94, cited in T. Kinnunen, H. S. Zamansky, and M. L. Block,

"Is the Hypnotized Subject Lying?," *Journal of Abnormal Psychology* 103, no. 2 (May 1994): 184–91.

14. Kinnunen, Zamansky, and Block, "Hypnotized Subject."

15. I. Kirsch, C. E. Silva, J. E. Carone, J. D. Johnston, and B. Simon, "The Surreptitious Observation Design: An Experimental Paradigm for Distinguishing Artifact from Essence in Hypnosis," *Journal of Abnormal Psychology* 98, no. 2 (May 1989): 132–36.

16. Orne and Evans, "Social Control."

17. W. C. Coe, K. Kobayashi, and M. L. Howard, "Experimental and Ethical Problems of Evaluating Influence of Hypnosis in Antisocial Conduct," *Journal of Abnormal Psychology* 82, no. 3 (December 1973): 476–82.

18. Coe, Kobayashi, and Howard, "Experimental and Ethical Problems," 481.

19. S. Milgram, "Behavioral Study of Obedience," *Journal of Abnormal and Social Psychology* 67, no. 4 (October 1963): 371–78.

20. N. S. Ward, D. A. Oakley, R. S. J. Frackowiak, and P. W. Halligan, "Differential Brain Activations during Intentionally Simulated and Subjectively Experienced Paralysis," *Cognitive Neuropsychiatry* 8, no. 4 (2003): 295–312.

21. H. Szechtman, E. Woody, K. S. Bowers, and C. Nahmias, "Where the Imaginal Appears Real: A Positron Emission Tomography Study of Auditory Hallucinations," *Proceedings of the National Academy of Sciences of the United States of America* 95, no. 4 (February 17, 1998): 1956–60.

22. S. M. Kosslyn, W. L. Thompson, M. F. Costantini-Ferrando, N. M. Alpert, and D. Spiegel, "Hypnotic Visual Illusion Alters Color Processing in the Brain," *American Journal of Psychiatry* 157, no. 8 (August 2000): 1279–84.

23. A. Raz, T. Shapiro, J. Fan, and M. I. Posner, "Hypnotic Suggestion and the Modulation of Stroop Interference," *Archives of General Psychiatry* 59, no. 12 (December 2002): 1155–61.

24. Raz et al., "Hypnotic Suggestion and Modulation."

25. A. Raz, J. Fan, and M. I. Posner, "Hypnotic Suggestion Reduces Conflict in the Human Brain," *Proceedings of the National Academy of Sciences of the United States of America* 102, no. 28 (July 12, 2005): 9978–9983.

CHAPTER 9: APPLIED MAGIC

1. T. Standage, *The Turk: The Life and Times of the Famous Eighteenth-Century Chess-Playing Machine* (New York: Berkley Books, 2003), 226.

2. D. Nuñez, M. Tempest, E. Viola, and C. Breazeal, "An Initial Discussion of Timing Considerations Raised during Development of a Magician-Robot Interaction" (workshop, 9th ACM/IEEE International Conference on Human-Robot Interaction, Bielefeld University, Bielefeld, Germany, March 3–6, 2014).

3. S. L'Yi, K. Koh, J. Jo, B. Kim, and J. Seo, "CloakingNote: A Novel Desktop Interface for Subtle Writing Using Decoy Texts," in *Proceedings of the 29th*

Annual Symposium on User Interface Software and Technology (New York: ACM, 2016), 473–481.

4. A. C. Clarke, *Profiles of the Future: An Inquiry into the Limits of the Possible* (London: Gollancz, 1999), 2.

5. B. Tognazzini, "Principles, Techniques, and Ethics of Stage Magic and Their Application to Human Interface Design," in *Proceedings of the INTERACT '93 and CHI '93 Conference on Human Factors in Computing Systems* (New York: ACM, 1993), 355–362.

6. M. Luo, "For Exercise in New York Futility, Push Button," *New York Times*, February 27, 2004, https://www.nytimes.com/2004/02/27/nyregion/for-exercise -in-new-york-futility-push-button.html.

7. E. Adar, D. S. Tan, and J. Teevan, "Benevolent Deception in Human Computer Interaction," in *Proceedings of the SIGCHI Conference on Human Factors in Computing Systems* (New York: ACM, 2013), 1863–1872.

8. Adar, D. S. Tan, and J. Teevan, "Benevolent Deception."

9. R. Hyman, "The Psychology of Deception," *Annual Review of Psychology* 40 (1989): 133–154.

10. See W. B. Yeager, *Techniques of the Professional Pickpocket* (Port Townsend, WA: Loompanics Unlimited, 1990); and S. R. Erdnase, *The Expert at the Card Table* (Toronto: Coles, 1980).

11. S. Rainey, "Derren Brown: 'Most Magicians Are Kleptomaniacs,'" *Telegraph*, May 26, 2013, https://www.telegraph.co.uk/culture/10081517/Derren-Brown -Most-magicians-are-kleptomaniacs.html.

12. J. E. Robert-Houdin, *Memoirs of Robert-Houdin: Ambassador, Author, and Conjuror* (London: Chapman and Hall, 1859), #:309–332.

13. J. Allen, "Deceptionists at War," *Cabinet*, no. 26 (Summer 2007).

14. See D. Fisher, *The War Magician* (New York: Berkley Books, 1983); and Allen, "Deceptionists at War."

15. See H. K. Melton and R. Wallace, *The Official CIA Manual of Trickery and Deception* (New York: William Morrow, 2009).

16. S. Henderson, R. Hoffman, L. Bunch, and J. Bradshaw, "Applying the Principles of Magic and the Concepts of Macrocognition to Counter-Deception in Cyber Operations" (paper presentation, International Naturalistic Decision Making Conference, McLean, VA, June 9–12, 2015).

17. W. R. Freudenburg and M. Alario, "Weapons of Mass Distraction: Magicianship, Misdirection, and the Dark Side of Legitimation," *Sociological Forum* 22, no. 2 (June 2007): 146–73.

18. Quoted in S. Delaney, "How Lynton Crosby (and a Dead Cat) Won the Election: 'Labour Were Intellectually Lazy,'" *Guardian*, January 20, 2016, https:// www.theguardian.com/politics/2016/jan/20/lynton-crosby-and-dead-cat-won -election-conservatives-labour-intellectually-lazy.

19. M. Solomon, "Up-to-Date Magic: Theatrical Conjuring and the Trick Film," *Theatre Journal* 58, no. 4 (December 2006): 595–615.

20. Solomon, "Up-to-Date Magic."

21. J. Steinmeyer, *The Last Greatest Magician in the World: Howard Thurston versus Houdini & the Battles of the American Wizards* (New York: Jeremy P. Tarcher/Penguin, 2012), 315.

22. "Paul Kieve Talks about Special Effects and Magic of Ghost," *Scotsman*, May 14, 2013, https://www.scotsman.com/lifestyle/culture/theatre/paul-kieve-talks -about-special-effects-and-magic-of-ghost-1-2927866.

23. S. Kumari, S. Deterding, and G. Kuhn, "Why Game Designers Should Study Magic," paper presented at Foundations of Digital Games 2018 (FDG18), August 7–10, 2018, Malmö, Sweden, https://doi.org/10.1145/3235765.3235788.

24. R. Manthorpe, "Derren Brown's VR Ghost Train Is Back—and This Time It's Actually Scary," *Wired UK*, March 31, 2017, http://www.wired.co.uk/article/ derren-brown-vr-ghost-train-thorpe-park.

25. S. H. Sharpe, *Conjurers' Optical Secrets* (Calgary: Hades, 1985), 11.

26. B. Blankenbehler, "How Greek Temples Correct Visual Distortion," *Architecture Revived*, October 15, 2015, http://www.architecturerevived.com/how-greek -temples-correct-visual-distortion/.

27. "The Magic of Architecture," SMP Architecture blog, October 12, 2016, http://www.sheerr.com/the-magic-of-architecture/.

28. S. Bagienski and G. Kuhn, "The Crossroads of Magic and Wellbeing: A Review of Wellbeing-Focused Magic Programs, Empirical Studies, and Conceivable Theories" (unpublished manuscript).

29. P. L. Harris, "Unexpected, Impossible and Magical Events: Children's Reactions to Causal Violations," *British Journal of Developmental Psychology* 12, no. 1 (March 1994): 1–7.

30. E. Subbotsky, "Curiosity and Exploratory Behaviour towards Possible and Impossible Events in Children and Adults," *British Journal of Psychology*, 101, no 3 (August 2010): 481–501.

31. E. Subbotsky, C. Hysted, and N. Jones, "Watching Films with Magical Content Facilitates Creativity in Children," *Perceptual and Motor Skills* 111, no. 1 (August 2010): 261–77.

32. Illusioneering, accessed December 14, 2017, http://www.illusioneering.org.

33. C. S. Hilas and A. Politis, "Motivating Students' Participation in a Computer Networks Course by Means of Magic, Drama and Games," *SpringerPlus* 3 (December 2014): 362.

34. M. Pritchard, "The Quantum State of Wonder," TEDxBrum, Birmingham, UK, October 2017, YouTube video, 8:19, November 16, 2017, https://www .youtube.com/watch?v=yVTBby0Dcg0.

35. S. A. Moss, M. Irons, and M. Boland, "The Magic of Magic: The Effect of Magic Tricks on Subsequent Engagement with Lecture Material," *British Journal*

of Educational Psychology 87, no. 1 (March 2017): 32–42; and B. A. Parris, G. Kuhn, G. A. Mizon, A. Benattayallah, and T. L. Hodgson, "Imaging the Impossible: An fMRI Study of Impossible Causal Relationships in Magic Tricks," *Neuro-Image* 45, no. 3 (April 15, 2009): 1033–39.

36. G. Labrocca and E. O. Piacentini, "Efficacia dei giochi di magia sul dolore da venipuntura: studio quasi sperimentale," *Children's Nurses: Italian Journal of Pediatric Nursing Science* 7, no. 1 (Primavera 2015): 4–5.

37. L. Vagnoli, S. Caprilli, A. Robiglio, and A. Messeri, "Clown Doctors as a Treatment for Preoperative Anxiety in Children: A Randomized, Prospective Study," *Pediatrics* 116, no. 4 (October 2005): e563–67.

38. B. Peretz and G. Gluck, "Magic Trick: A Behavioural Strategy for the Management of Strong-Willed Children," *International Journal of Paediatric Dentistry* 15, no. 6 (November 2005): 429–36.

39. D. Green, M. Schertz, A. M. Gordon, A. Moore, T. Schejter Margalit, Y. Farquharson, D. Ben Bashat, et al., "A Multi-site Study of Functional Outcomes Following a Themed Approach to Hand-Arm Bimanual Intensive Therapy for Children with Hemiplegia," *Developmental Medicine & Child Neurology* 55, no. 6 (June 2013): 527–33.

40. K. Spencer, "Hocus Focus: Evaluating the Academic and Functional Benefits of Integrating Magic Tricks in the Classroom," *Journal of the International Association of Special Education* 13, no. 1 (Spring 2012): 87–99.

41. Abracademy, accessed December 14, 2017, https://www.abracademy.com.

42. Streets of Growth, accessed December 14, 2017, http://www.streetsofgrowth.org.

43. S. Lachapelle, "From the Stage to the Laboratory: Magicians, Psychologists, and the Science of Illusion," *Journal of the History of the Behavioral Sciences* 44, no. 4 (Autumn 2008): 319–34.

44. M. Faraday, "Experimental Investigation of Table-Moving," *Athenæum, Journal of English Foreign Literature, Science, and the Fine Arts*, no. 1340 (July 2, 1853): 801–3; and R. Hodgson and S. J. Davey, "The Possibilities of Malobservation and Lapse of Memory from a Practical Point of View," *Proceedings of the Society for Psychical Research* 4 (1886–87): 381–495.

45. M. Dessoir, "The Psychology of Legerdemain," *The Open Court* 12 (1893): 3599–3606.

46. Dessoir, "The Psychology of Legerdemain."

47. See G. Kuhn and B. W. Tatler, "Magic and Fixation: Now You Don't See It, Now You Do," *Perception* 34, no. 9 (September 2005): 1155–61.

48. See G. Kuhn and B. W. Tatler, "Misdirected by the Gap: The Relationship between Inattentional Blindness and Attentional Misdirection," *Consciousness and Cognition* 20, no. 2 (June 2011): 432–36.

49. Dessoir, "The Psychology of Legerdemain."

50. See J. A. Olson, A. A. Amlani, A. Raz, and R. A. Rensink, "Influencing Choice without Awareness," *Consciousness and Cognition* 37 (December 2015): 225–36.

51. A. Binet, "La psychologie de la prestidigitation," *Revue des Deux Mondes* 125 (October 15, 1894): 903–22.

52. C. Thomas, A. Didierjean, and S. Nicolas, "Scientific Study of Magic: Binet's Pioneering Approach Based on Observations and Chronophotography," *American Journal of Psychology* 129, no. 3 (September 2016): 313–26.

53. Binet, "Psychologie de la prestidigitation."

54. J. Jastrow, "Psychological Notes upon Sleight-of-Hand Experts," *Science* 71, no. 3 (May 8, 1896): 685–89.

55. N. Triplett, "The Psychology of Conjuring Deceptions," *American Journal of Psychology* 11, no. 4 (July 1900): 439–510.

56. Hyman, "Psychology of Deception."

57. P. Lamont and R. Wiseman, *Magic in Theory: An Introduction to the Theoretical and Psychological Elements of Conjuring* (Seattle: Hermetic Press, 1999).

58. G. Kuhn, A. A. Amlani, and R. A. Rensink, "Toward a Science of Magic," *Trends in Cognitive Sciences* 12, no. 9 (September 2008): 349–54.

59. S. L. Macknik, M. King, J. Randi, A. Robbins, Teller, J. Thompson, and S. Martinez-Conde, "Attention and Awareness in Stage Magic: Turning Tricks into Research," *Nature Reviews Neuroscience* 9, no. 11 (November 2008): 871–79.

60. S. L. Macknik, S. Martinez-Conde, and S. Blakeslee, *Sleights of Mind: What the Neuroscience of Magic Reveals about Our Everyday Deceptions* (New York: Henry Holt, 2010).

61. P. Lamont, J. M. Henderson, and T. J. Smith, "Where Science and Magic Meet: The Illusion of a 'Science of Magic,'" *Review of General Psychology* 14, no. 1 (March 2010): 16–21.

62. A famous example of this is the study of neural networks, which resumed twenty years after work on the topic had been deemed pointless. Neural networks are now one of the biggest success stories in the field of artificial intelligence.

63. See F. Phillips, M. B. Natter, and E. J. L. Egan, "Magically Deceptive Biological Motion—The French Drop Sleight," *Frontiers in Psychology*, April 9, 2015; and C. Cavina-Pratesi, G. Kuhn, M. Ietswaart, and A. D. Milner, "The Magic Grasp: Motor Expertise in Deception," *PLoS ONE* 6, no. 2 (2011): e16568.

64. See G. Kuhn and L. M. Martinez, "Misdirection—Past, Present, and the Future," *Frontiers in Human Neuroscience*, January 6, 2012.

65. See Parris et al., "Imaging the Impossible"; A. H. Danek, M. Öllinger, T. Fraps, B. Grothe, and V. L. Flanagin, "An fMRI Investigation of Expectation Violation in Magic Tricks," *Frontiers in Psychology*, February 4, 2015; and H. A. Caffaratti, J. Navajas, H. G. Rey, and R. Quian Quiroga, "Where Is the Ball? Behavioral and Neural Responses Elicited by a Magic Trick," *Psychophysiology* 53, no. 9 (September 2016): 1441–48.

66. See R. A. Rensink and G. Kuhn, "A Framework for Using Magic to Study the Mind," *Frontiers in Psychology*, February 2, 2015.

67. R. Baillargeon and J. DeVos, "Object Permanence in Young Infants: Further Evidence," *Child Development* 62, no. 6 (December 1991): 1227–46.

68. See G. Kuhn, and R. Teszka, "Attention and Misdirection: How to Use Conjuring Experience to Study Attentional Processes," *Handbook of Attention*, ed. J. M. Fawcett, E. F. Risko, and A. Kingstone (Cambridge, MA: MIT Press, 2016), 503–24; P. Johansson, L. Hall, S. Sikstrom, and A. Olsson, "Failure to Detect Mismatches between Intention and Outcome in a Simple Decision Task," *Science* 310, no. 5745 (October 7, 2005): 116–19; and B. M. Hood and P. Bloom, "Children Prefer Certain Individuals over Perfect Duplicates," *Cognition* 106, no. 1 (January 2008): 455–62.

69. See A. H. Danek, T. Fraps, A. von Müller, B. Grothe, and M. Öllinger, "Working Wonders? Investigating Insight with Magic Tricks," *Cognition* 130, no. 2 (February 2014): 174–85.

70. See C. Mohr, N. Koutrakis, and G. Kuhn, "Priming Psychic and Conjuring Abilities of a Magic Demonstration Influences Event Interpretation and Random Number Generation Biases," January 21, 2015, January 21, 2015; E. Subbotsky, "Can Magical Intervention Affect Subjective Experiences? Adults' Reactions to Magical Suggestion," *British Journal of Psychology* 100, no. 3 (August 2009): 517–537; and L. Lesaffre, C. Mohr, D. Rochat, G. Kuhn, and A. Abu-Akel, "Effects of Magic Performances on Belief and Random Number Generation" (unpublished manuscript).

71. See V. Ekroll, B. Sayim, and J. Wagemans, "The Other Side of Magic: The Psychology of Perceiving Hidden Things," *Perspectives on Psychological Science* 12, no. 1 (January 2017): 91–106.

72. See G. Kuhn, H. A. Caffaratti, R. Teszka, and R. A. Rensink, "A Psychologically-Based Taxonomy of Misdirection," *Frontiers in Psychology*, December 9, 2014, doi:10.3389/fpsyg.2014.01392.

73. See Olson et al., "Influencing Choice."

74. C. Thomas and A. Didierjean, "Magicians Fix Your Mind: How Unlikely Solutions Block Obvious Ones," *Cognition* 154 (September 2016): 169–173; and C. Thomas, A. Didierjean, and G. Kuhn, "It Is Magic! How Impossible Solutions Prevent the Discovery of Obvious Ones?," *Quarterly Journal of Experimental Psychology*, published online ahead of print on January 1, 2018, doi:10.1177/1747021817743439.

75. C. Thomas, A. Didierjean, and G. Kuhn, "The Flushtration Count Illusion: Attribute Substitution Tricks Our Interpretation of a Simple Visual Event Sequence," *British Journal of Psychology*, published online ahead of print on April 17, 2018, doi:10.1111/bjop.12306.

76. Rensink and Kuhn, "Framework for Using Magic."

77. P. Lamont, "Problems with the Mapping of Magic Tricks," *Frontiers in Psychology*, June 23, 2015; and R. A. Rensink and G. Kuhn, "The Possibility of a Science of Magic," *Frontiers in Psychology*, October 16, 2015.

78. Kuhn et al., "Psychologically-Based Taxonomy."

79. M. W. Eysenck and M. T. Keane, *Cognitive Psychology: A Student's Hand-book*, 7th ed. (London: Psychology Press, 2015).

80. Rensink and Kuhn, "Framework for Using Magic."

CHAPTER 10: HOW TO ADVANCE THE MAGIC ENDEAVOR?

1. J. E. Robert-Houdin, *The Secrets of Conjuring and Magic*, trans. and ed. Professor Hoffman (London: George Routledge and Sons, 1878), 29.

2. J. Tamariz, "Fundamentals in Illusionism," in *Magic in Mind: Essential Essays for Magicians*, ed. J. Jay (n.p.: Vanishing Inc. Magic Shop, 2012), 59

3. See G. G. Cole, P. A. Skarratt, and G. Kuhn, "Real Person Interaction in Visual Attention Research," *European Psychologist* 21, no. 2 (2016): 141–49.

4. See A. Binet, "Psychology of Prestidigitation," in *Annual Report of the Board of Regents of the Smithsonian Institution, Showing the Operations, Expenditures, and Conditions of the Institution to July, 1894* (Washington, DC: Government Printing Office, 1896), 555–71; G. Cocchini, T. Galligan, L. Mora, and G. Kuhn, "The Magic Hand: Plasticity of Mental Hand Representation," *Quarterly Journal of Experimental Psychology*, January 1, 2018; and C. Cavina-Pratesi, G. Kuhn, M. Ietswaart, and A. D. Milner, "The Magic Grasp: Motor Expertise in Deception," *PLoS ONE* 6, no. 2 (2011): e16568.

5. B. A. Parris, G. Kuhn, G. A. Mizon, A. Benattayallah, and T. L. Hodgson, " Imaging the Impossible: An fMRI Study of Impossible Causal Relationships in Magic Tricks," *NeuroImage* 45, no. 3 (April 15, 2009): 1033–39.

6. B. Cox and R. Ince, "Christmas Special: The Science of Magic," December 25, 2017, in *The Infinite Monkey Cage*, produced by A. Feachem, BBC Radio 4, 27:46, https://www.bbc.co.uk/programmes/b09jrtb6.

7. See R. A. Rensink, J. K. O'Regan, and J. J. Clark, "To See or Not to See: The Need for Attention to Perceive Changes in Scenes," *Psychological Science* 8, no. 5 (September 1997): 368–73.

8. R. Wiseman, "Color Changing Card Trick," YouTube video, 2:43, November 21, 2012, https://www.youtube.com/watch?v=v3iPrBrGSJM.

9. J. Lehrer, "Magic and the Brain: Teller Reveals the Neuroscience of Illusion," *Wired*, April 20, 2009, https://www.wired.com/2009/04/ff-neuroscienceofmagic/; and T. Stone, "Cognitive Conjuring," Lodestones, *Genii*, 74, no. 3 (March 2011): 40–43.

10. P. Lamont and R. Wiseman, *Magic in Theory: An Introduction to the Theoretical and Psychological Elements of Conjuring* (Seattle: Hermetic Press, 1999), 37; and G. Kuhn and L. M. Martinez, "Misdirection—Past, Present, and the Future," *Frontiers in Human Neuroscience*, January 6, 2012.

11. G. Kuhn, R. Teszka, N. Tenaw, and A. Kingstone, "Don't Be Fooled! Attentional Responses to Social Cues in a Face-to-Face and Video Magic Trick Reveals Greater Top-Down Control for Overt than Covert Attention," *Cognition* 146 (January 2016): 136–42.

12. D. J. Simons and C. F. Chabris, "Gorillas in Our Midst: Sustained Inattentional Blindness for Dynamic Events," *Perception* 28, no. 9 (September 1999): 1059–74.

13. C. N. L. Olivers, F. Meijer, and J. Theeuwes, "Feature-Based Memory-Driven Attentional Capture: Visual Working Memory Content Affects Visual Attention," *Journal of Experimental Psychology: Human Perception and Performance* 32, no. 5 (October 2006): 1243.

14. D. E. Shalom, M. G. de Sousa Serro, M. Giaconia, L. M. Martinez, A. Rieznik, and M. Sigman, "Choosing in Freedom or Forced to Choose? Introspective Blindness to Psychological Forcing in Stage-Magic," *PloS ONE* 8, no. 3 (2013): e58254.

15. O. Rissanen, P. Pitkänen, A. Juvonen, G. Kuhn, and K. Hakkarainen, "Expertise among Professional Magicians: An Interview Study," *Frontiers in Psychology*, December 23, 2014.

16. A. H. Danek, M. Öllinger, T. Fraps, B. Grothe, and V. L. Flanagin, "An fMRI Investigation of Expectation Violation in Magic Tricks," *Frontiers in Psychology*, February 4, 2015.

17. "The Importance of Combining Methods," *Jerx* (blog), July 1, 2016, http://www.thejerx.com/blog/2016/6/30/the-importance-of-combining-methods.

18. J. Jay, "What Do Audiences Really Think?," *Magic*, September 2016, 46–55.

19. "The Importance of Combining Methods."

20. H. Williams and P. W. McOwan, "Manufacturing Magic and Computational Creativity," *Frontiers in Psychology*, June 10, 2016; and H. Williams and P. W. McOwan, "Magic in the Machine: A Computational Magician's Assistant," *Frontiers in Psychology*, November 17, 2014.

21. H. Williams, "The Twelve Magicians of Osiris," YouTube video, 0:41, November 4, 2014, https://www.youtube.com/watch?v=PsBWBs1pIG0.

22. J. A. Olson, A. A. Amlani, and R. A. Rensink, "Perceptual and Cognitive Characteristics of Common Playing Cards," *Perception* 41, no. 3 (March 2012): 268–86.

23. J. Leddington, "The Enjoyment of Negative Emotions in the Experience of Magic," *Behavioral and Brain Sciences* 40 (2017): e369.

24. J. Leddington, "The Experience of Magic," *Journal of Aesthetics and Art Criticism* 74, no. 3 (Summer 2016): 253–64.

25. J. I. Swiss, *Shattering Illusions: Essays on the Ethics, History, and Presentation of Magic* (Seattle: Hermetic Press, 2002), 6.

26. Leddington, "Enjoyment of Negative Emotions."

27. D. Ortiz, *Strong Magic* (n.p.: Ortiz Publications, 1994), 30.

28. W. Menninghaus, V. Wagner, J. Hanich, E. Wassiliwizky, T. Jacobsen, and S. Koelsch, "The Distancing–Embracing Model of the Enjoyment of Negative Emotions in Art Reception," *Behavioral and Brain Sciences* 40 (2017): e347.

29. E. B. Andrade and J. B. Cohen, "On the Consumption of Negative Feelings," *Journal of Consumer Research* 34, no. 3 (October 2007): 283–300.

30. Menninghaus et al., "The Distancing–Embracing Model."

31. Menninghaus et al., "The Distancing–Embracing Model."

32. Leddington, "Enjoyment of Negative Emotions."

33. Jay, "What Do Audiences Really Think?"

34. T. L. Griffiths, "Revealing Ontological Commitments by Magic," *Cognition* 136 (March 2015): 43–48.

35. E. Subbotsky, *Magic and the Mind: Mechanisms, Functions, and Development of Magical Thinking and Behavior* (Oxford: Oxford University Press, 2010), 165.

36. L. Lesaffre, C. Mohr, D. Rochat, G. Kuhn, and A. Abu-Akel, "Magic Performances—When Explained in Psychic Terms by University Students" (unpublished manuscript).

37. Lesaffre et al., "Effects of Magic Performances."

38. Jay, "What Do Audiences Really Think?"

39. Jay, "What Do Audiences Really Think?"

40. See P. M. Nardi, "The Social World of Magicians: Gender and Conjuring," *Sex Roles* 19, no. 11–12 (December 1988): 759–70; G. M. Jones, *Trade of the Tricks: Inside the Magician's Craft* (Berkeley: University of California Press, 2011); and Rissanen et al., "Expertise among Professional Magicians."

41. A. Fetters, "Why Are There So Few Female Magicians?," *Atlantic*, March 18, 2013, https://www.theatlantic.com/entertainment/archive/2013/03/why-are -there-so-few-female-magicians/274099/.

42. Nardi, "Social World of Magicians."

43. P. Lamont, "Modern Magic, the Illusion of Transformation, and How It Was Done," *Journal of Social History*, November 25, 2016, shw126.

44. Nardi, "The Social World of Magicians."

45. L. C. Bruns and J. P. Zompetti, "The Rhetorical Goddess: A Feminist Perspective on Women in Magic," *Journal of Performance Magic* 2, no. 1 (2014): 8–39.

46. H. Nelms, "Consistency in Characterization," in *Magic and Showmanship: A Handbook for Conjurers* (New York: Dover, 1969), 63–75.

47. Bruns and Zompetti, "The Rhetorical Goddess."

48. Nardi, "The Social World of Magicians."

49. O. Rissanen, P. Pitkänen, A. Juvonen, P. Räihä, G. Kuhn, and K. Hakkarainen, "How Has the Emergence of Digital Culture Affected Professional Magic?," *Professions and Professionalism* 7, no. 3 (2017): e1957.

50. Nardi, "The Social World of Magicians," 24.

51. Nardi, "The Social World of Magicians," 24.

52. Bruns and Zompetti, "The Rhetorical Goddess."

53. P. Gygax, "When Sexism Creeps into Perception: How Gender Stereotypes Affect the Way We Perceive a Magic Trick" (paper presentation, Science of Magic Association Conference, Goldsmiths, University of London, September 1, 2017).

54. S. Lachapelle, "From the Stage to the Laboratory: Magicians, Psychologists, and the Science of Illusion," *Journal of the History of the Behavioral Sciences* 44, no. 4 (Autumn 2008): 319–34.

CHAPTER 11: CONCLUSION

1. O. Rissanen, P. Pitkänen, A. Juvonen, P. Räihä, G. Kuhn, and K. Hakkarainen, "How Has the Emergence of Digital Culture Affected Professional Magic?," *Professions and Professionalism* 7, no. 3 (2017): e1957.

INDEX